MACHINE LEARNING Q AND AI

MACHINE LEARNING Q AND AI

30 Essential Questions and Answers on Machine Learning and AI

by Sebastian Raschka

no starch press®

San Francisco

Printed in the United States of America

First printing

28 27 26 25 24   1 2 3 4 5

ISBN-13: 978-1-7185-0376-2 (print)
ISBN-13: 978-1-7185-0377-9 (ebook)

 Published by No Starch Press®, Inc.
245 8th Street, San Francisco, CA 94103
phone: +1.415.863.9900
www.nostarch.com; info@nostarch.com

Publisher: William Pollock
Managing Editor: Jill Franklin
Production Manager: Sabrina Plomitallo-González
Production Editor: Miles Bond
Developmental Editor: Abigail Schott-Rosenfield
Cover Illustrator: Gina Redman
Interior Design: Octopod Studios
Technical Reviewer: Andrea Panizza
Copyeditor: George Hale
Proofreader: Audrey Doyle
Indexer: BIM Creatives, LLC

Library of Congress Control Number: 2023041820

For customer service inquiries, please contact info@nostarch.com. For information on distribution, bulk sales, corporate sales, or translations: sales@nostarch.com. For permission to translate this work: rights@nostarch.com. To report counterfeit copies or piracy: counterfeit@nostarch.com.

[S]

To my partner, Liza; my family; and the global community of creators who have motivated and influenced my journey as an author

About the Author

Sebastian Raschka, PhD, is a machine learning and AI researcher with a strong passion for education. As Lead AI Educator at Lightning AI, he is excited about making AI and deep learning more accessible and teaching people how to utilize these technologies at scale. Before dedicating his time fully to Lightning AI, Sebastian held a position as assistant professor of statistics at the University of Wisconsin–Madison, where he specialized in researching deep learning and machine learning. You can find out more about his research on his website (*https://sebastianraschka.com*). Moreover, Sebastian loves open source software and has been a passionate contributor for over a decade. Next to coding, he also loves writing and authored the bestselling books *Python Machine Learning* and *Machine Learning with PyTorch and Scikit-Learn* (both from Packt Publishing).

About the Technical Reviewer

Andrea Panizza is a principal AI specialist at Baker Hughes, leveraging cutting-edge AI/ML techniques to accelerate engineering design, automate information retrieval and extraction from large collections of documents, and use computer vision to power unmanned asset inspection. He has a PhD in computational fluid dynamics. Before joining Baker Hughes, he worked as a CFD researcher at CIRA (the Italian Center for Aerospace Research).

BRIEF CONTENTS

PART IV: PRODUCTION AND DEPLOYMENT

PART V: PREDICTIVE PERFORMANCE AND MODEL EVALUATION

CONTENTS IN DETAIL

PART I
NEURAL NETWORKS AND DEEP LEARNING

8
THE SUCCESS OF TRANSFORMERS　　　　　　　　　　　　43

9
GENERATIVE AI MODELS　　　　　　　　　　　　　　　　49

10
SOURCES OF RANDOMNESS　　　　　　　　　　　　　59

PART II
COMPUTER VISION

PART III
NATURAL LANGUAGE PROCESSING

PART V
PREDICTIVE PERFORMANCE AND MODEL EVALUATION

24
POISSON AND ORDINAL REGRESSION   161

25
CONFIDENCE INTERVALS   163

26
CONFIDENCE INTERVALS VS. CONFORMAL PREDICTIONS   173

27
PROPER METRICS   179

FOREWORD

There are hundreds of introductory texts on machine learning. They come in a vast array of styles and approaches, from theory-first perspectives for graduate students to business-focused viewpoints for C-suites. These primers are invaluable resources for individuals taking their first steps into the field, and they will remain so for decades to come.

However, the path to expertise is not solely about the beginnings. It also encompasses the intricate detours, the steep ascents, and the nuances that aren't evident initially. Put another way, after studying the basics, learners are left asking, "What's next?" It is here, in the realm beyond the basics, that this book finds its purpose.

Within these pages, Sebastian walks readers through a broad spectrum of intermediate and advanced topics in applied machine learning they are likely to encounter on their journey to expertise. One could hardly ask for a better guide than Sebastian, who is, without exaggeration, the best machine learning educator currently in the field. On each page, Sebastian not only imparts his extensive knowledge, but also shares the passion and curiosity that mark true expertise.

To all the learners who have crossed the initial threshold and are eager to delve deeper, this book is for you. You will finish this book as a more skilled machine learning practitioner than when you began. Let it serve as the bridge to your next phase of rewarding adventures in machine learning.

Good luck.

Chris Albon
Director of Machine Learning, the Wikimedia Foundation
San Francisco
August 2023

ACKNOWLEDGMENTS

Writing a book is an enormous undertaking. This project would not have been possible without the help of the open source and machine learning communities who collectively created the technologies that this book is about.

I want to extend my thanks to the following people for their immensely helpful feedback on the manuscript:

- Andrea Panizza for being an outstanding technical reviewer, who provided very valuable and insightful feedback.

- Anton Reshetnikov for suggesting a cleaner layout for the supervised learning flowchart in Chapter 30.

- Nikan Doosti, Juan M. Bello-Rivas, and Ken Hoffman for sharing various typographical errors.

- Abigail Schott-Rosenfield and Jill Franklin for being exemplary editors. Their knack for asking the right questions and improving language has significantly elevated the quality of this book.

INTRODUCTION

Thanks to rapid advancements in deep learning, we have seen a significant expansion of machine learning and AI in recent years.

This progress is exciting if we expect these advancements to create new industries, transform existing ones, and improve the quality of life for people around the world. On the other hand, the constant emergence of new techniques can make it challenging and time-consuming to keep abreast of the latest developments. Nonetheless, staying current is essential for professionals and organizations that use these technologies.

I wrote this book as a resource for readers and machine learning practitioners who want to advance their expertise in the field and learn about techniques that I consider useful and significant but that are often overlooked in traditional and introductory textbooks and classes. I hope you'll find this book a valuable resource for obtaining new insights and discovering new techniques you can implement in your work.

Who Is This Book For?

Navigating the world of AI and machine learning literature can often feel like walking a tightrope, with most books positioned at either end: broad beginner's introductions or deeply mathematical treatises. This book illustrates and discusses important developments in these fields while staying approachable and not requiring an advanced math or coding background.

This book is for people with some experience with machine learning who want to learn new concepts and techniques. It's ideal for those who have taken a beginner course in machine learning or deep learning or have read an equivalent introductory book on the topic. (Throughout this book, I will use *machine learning* as an umbrella term for machine learning, deep learning, and AI.)

What Will You Get Out of This Book?

This book adopts a unique Q&A style, where each brief chapter is structured around a central question related to fundamental concepts in machine learning, deep learning, and AI. Every question is followed by an explanation, with several illustrations and figures, as well as exercises to test your understanding. Many chapters also include references for further reading. These bite-sized nuggets of information provide an enjoyable jumping-off point on your journey from machine learning beginner to expert.

The book covers a wide range of topics. It includes new insights about established architectures, such as convolutional networks, that allow you to utilize these technologies more effectively. It also discusses more advanced techniques, such as the inner workings of large language models (LLMs) and vision transformers. Even experienced machine learning researchers and practitioners will encounter something new to add to their arsenal of techniques.

While this book will expose you to new concepts and ideas, it's not a math or coding book. You won't need to solve any proofs or run any code while reading. In other words, this book is a perfect travel companion or something you can read on your favorite reading chair with your morning coffee or tea.

How to Read This Book

Each chapter of this book is designed to be self-contained, offering you the freedom to jump between topics as you wish. When a concept from one chapter is explained in more detail in another, I've included chapter references you can follow to fill in gaps in your understanding.

However, there's a strategic sequence to the chapters. For example, the early chapter on embeddings sets the stage for later discussions on self-supervised learning and few-shot learning. For the easiest reading experience and the most comprehensive grasp of the content, my recommendation is to approach the book from start to finish.

Each chapter is accompanied by optional exercises for readers who want to test their understanding, with an answer key located at the end of the book. In addition, for any papers referenced in a chapter or further reading on that chapter's topic, you can find the complete citation information in that chapter's "References" section.

The book is structured into five main parts centered on the most important topics in machine learning and AI today.

Part I: Neural Networks and Deep Learning covers questions about deep neural networks and deep learning that are not specific to a particular subdomain. For example, we discuss alternatives to supervised learning and techniques for reducing overfitting, which is a common problem when using machine learning models for real-world problems where data is limited.

Chapter 1: Embeddings, Latent Space, and Representations   Delves into the distinctions and similarities between embedding vectors, latent vectors, and representations. Elucidates how these concepts help encode information in the context of machine learning.

Chapter 2: Self-Supervised Learning   Focuses on self-supervised learning, a method that allows neural networks to utilize large, unlabeled datasets in a supervised manner.

Chapter 3: Few-Shot Learning   Introduces few-shot learning, a specialized supervised learning technique tailored for small training datasets.

Chapter 4: The Lottery Ticket Hypothesis   Explores the idea that randomly initialized neural networks contain smaller, efficient subnetworks.

Chapter 5: Reducing Overfitting with Data   Addresses the challenge of overfitting in machine learning, discussing strategies centered on data augmentation and the use of unlabeled data to reduce overfitting.

Chapter 6: Reducing Overfitting with Model Modifications   Extends the conversation on overfitting, focusing on model-related solutions like regularization, opting for simpler models, and ensemble techniques.

Chapter 7: Multi-GPU Training Paradigms   Explains various training paradigms for multi-GPU setups to accelerate model training, including data and model parallelism.

Chapter 8: The Success of Transformers   Explores the popular transformer architecture, highlighting features like attention mechanisms, parallelization ease, and high parameter counts.

Chapter 9: Generative AI Models   Provides a comprehensive overview of deep generative models, which are used to produce various media forms, including images, text, and audio. Discusses the strengths and weaknesses of each model type.

Chapter 10: Sources of Randomness   Addresses the various sources of randomness in the training of deep neural networks that may lead to inconsistent and non-reproducible results during both training and inference. While randomness can be accidental, it can also be intentionally introduced by design.

Part II: Computer Vision focuses on topics mainly related to deep learning but specific to computer vision, many of which cover convolutional neural networks and vision transformers.

Chapter 11: Calculating the Number of Parameters   Explains the procedure for determining the parameters in a convolutional neural network, which is useful for gauging a model's storage and memory requirements.

Chapter 12: Fully Connected and Convolutional Layers   Illustrates the circumstances in which convolutional layers can seamlessly replace fully connected layers, which can be useful for hardware optimization or simplifying implementations.

Chapter 13: Large Training Sets for Vision Transformers   Probes the rationale behind vision transformers requiring more extensive training sets compared to conventional convolutional neural networks.

Part III: Natural Language Processing covers topics around working with text, many of which are related to transformer architectures and self-attention.

Chapter 14: The Distributional Hypothesis   Delves into the distributional hypothesis, a linguistic theory suggesting that words appearing in the same contexts tend to possess similar meanings, which has useful implications for training machine learning models.

Chapter 15: Data Augmentation for Text   Highlights the significance of data augmentation for text, a technique used to artificially increase dataset sizes, which can help with improving model performance.

Chapter 16: Self-Attention   Introduces self-attention, a mechanism allowing each segment of a neural network's input to refer to other parts. Self-attention is a key mechanism in modern large language models.

Chapter 17: Encoder- and Decoder-Style Transformers   Describes the nuances of encoder and decoder transformer architectures and explains which type of architecture is most useful for each language processing task.

Chapter 18: Using and Fine-Tuning Pretrained Transformers   Explains different methods for fine-tuning pretrained large language models and discusses their strengths and weaknesses.

Chapter 19: Evaluating Generative Large Language Models   Lists prominent evaluation metrics for language models like Perplexity, BLEU, ROUGE, and BERTScore.

Part IV: Production and Deployment covers questions pertaining to practical scenarios, such as increasing inference speeds and various types of distribution shifts.

Chapter 20: Stateless and Stateful Training   Distinguishes between stateless and stateful training methodologies used in deploying models.

Chapter 21: Data-Centric AI   Explores data-centric AI, which prioritizes refining datasets to enhance model performance. This approach contrasts with the conventional model-centric approach, which emphasizes improving model architectures or methods.

Chapter 22: Speeding Up Inference   Introduces techniques to enhance the speed of model inference without tweaking the model's architecture or compromising accuracy.

Chapter 23: Data Distribution Shifts   Post-deployment, AI models may face discrepancies between training data and real-world data distributions, known as data distribution shifts. These shifts can deteriorate model performance. This chapter categorizes and elaborates on common shifts like covariate shift, concept drift, label shift, and domain shift.

Part V: Predictive Performance and Model Evaluation dives deeper into various aspects of squeezing out predictive performance, such as changing the loss function, setting up k-fold cross-validation, and dealing with limited labeled data.

Chapter 24: Poisson and Ordinal Regression   Highlights the differences between Poisson and ordinal regression. Poisson regression is suitable for count data that follows a Poisson distribution, like the number of colds contracted on an airplane. In contrast, ordinal regression caters to ordered categorical data without assuming equidistant categories, such as disease severity.

Chapter 25: Confidence Intervals   Delves into methods for constructing confidence intervals for machine learning classifiers. Reviews the purpose of confidence intervals, discusses how they estimate unknown population parameters, and introduces techniques such as normal approximation intervals, bootstrapping, and retraining with various random seeds.

Chapter 26: Confidence Intervals vs. Conformal Predictions   Discusses the distinction between confidence intervals and conformal predictions and describes the latter as a tool for creating prediction intervals that cover actual outcomes with specific probability.

Chapter 27: Proper Metrics   Focuses on the essential properties of a proper metric in mathematics and computer science. Examines whether commonly used loss functions in machine learning, such as mean squared error and cross-entropy loss, satisfy these properties.

Chapter 28: The k in k-Fold Cross-Validation   Explores the role of the k in k-fold cross-validation and provides insight into the advantages and disadvantages of selecting a large k.

Chapter 29: Training and Test Set Discordance   Addresses the scenario where a model performs better on a test dataset than the training dataset. Offers strategies to discover and address discrepancies between training and test datasets, introducing the concept of adversarial validation.

Chapter 30: Limited Labeled Data Introduces various techniques to enhance model performance in situations where data is limited. Covers data labeling, bootstrapping, and paradigms such as transfer learning, active learning, and multimodal learning.

Online Resources

I've provided optional supplementary materials on GitHub with code examples for certain chapters to enhance your learning experience (see *https://github.com/rasbt/MachineLearning-QandAI-book*). These materials are designed as practical extensions and deep dives into topics covered in the book. You can use them alongside each chapter or explore them after reading to solidify and expand your knowledge.

Without further ado, let's dive in.

PART I

NEURAL NETWORKS AND DEEP LEARNING

1

EMBEDDINGS, LATENT SPACE, AND REPRESENTATIONS

In deep learning, we often use the terms *embedding vectors*, *representations*, and *latent space*. What do these concepts have in common, and how do they differ?

While these three terms are often used interchangeably, we can make subtle distinctions between them:

- Embedding vectors are representations of input data where similar items are close to each other.
- Latent vectors are intermediate representations of input data.
- Representations are encoded versions of the original input.

The following sections explore the relationship between embeddings, latent vectors, and representations and how each functions to encode information in machine learning contexts.

Embeddings

Embedding vectors, or *embeddings* for short, encode relatively high-dimensional data into relatively low-dimensional vectors.

We can apply embedding methods to create a continuous dense (nonsparse) vector from a (sparse) one-hot encoding. *One-hot encoding* is a method used to represent categorical data as binary vectors, where each category is mapped to a vector containing 1 in the position corresponding to the category's index, and 0 in all other positions. This ensures that the categorical values are represented in a way that certain machine learning algorithms can process. For example, if we have a categorical variable Color with three categories, Red, Green, and Blue, the one-hot encoding would represent Red as [1, 0, 0], Green as [0, 1, 0], and Blue as [0, 0, 1]. These one-hot encoded categorical variables can then be mapped into continuous embedding vectors by utilizing the learned weight matrix of an embedding layer or module.

We can also use embedding methods for dense data such as images. For example, the last layers of a convolutional neural network may yield embedding vectors, as illustrated in Figure 1-1.

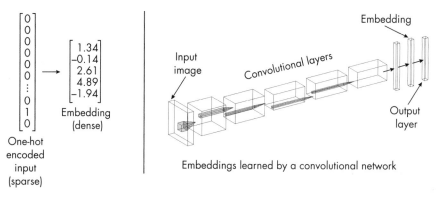

Figure 1-1: An input embedding (left) and an embedding from a neural network (right)

To be technically correct, all intermediate layer outputs of a neural network could yield embedding vectors. Depending on the training objective, the output layer may also produce useful embedding vectors. For the sake of simplicity, the convolutional neural network in Figure 1-1 associates the second-to-last layer with embeddings.

Embeddings can have higher or lower numbers of dimensions than the original input. For instance, using embeddings methods for extreme expression, we can encode data into two-dimensional dense and continuous representations for visualization purposes and clustering analysis, as illustrated in Figure 1-2.

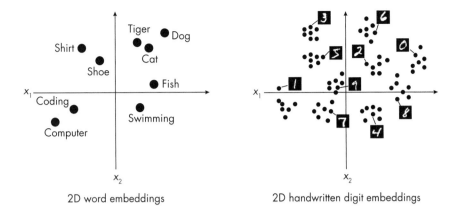

2D word embeddings 2D handwritten digit embeddings

Figure 1-2: Mapping words (left) and images (right) to a two-dimensional feature space

A fundamental property of embeddings is that they encode *distance* or *similarity*. This means that embeddings capture the semantics of the data such that similar inputs are close in the embeddings space.

For readers interested in a more formal explanation using mathematical terminology, an embedding is an injective and structure-preserving map between an input space X and the embedding space Y. This implies that similar inputs will be located at points in close proximity within the embedding space, which can be seen as the "structure-preserving" characteristic of the embedding.

Latent Space

Latent space is typically used synonymously with *embedding space*, the space into which embedding vectors are mapped.

Similar items can appear close in the latent space; however, this is not a strict requirement. More loosely, we can think of the latent space as any feature space that contains features, often compressed versions of the original input features. These latent space features can be learned by a neural network, such as an autoencoder that reconstructs input images, as shown in Figure 1-3.

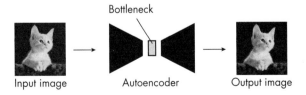

Input image Autoencoder Output image

Figure 1-3: An autoencoder reconstructing the input image

The bottleneck in Figure 1-3 represents a small, intermediate neural network layer that encodes or maps the input image into a lower-dimensional representation. We can think of the target space of this mapping as a latent space. The training objective of the autoencoder is to reconstruct the input image, that is, to minimize the distance between the input and output images. In order to optimize the training objective, the autoencoder may learn to place the encoded features of similar inputs (for example, pictures of cats) close to each other in the latent space, thus creating useful embedding vectors where similar inputs are close in the embedding (latent) space.

Representation

A *representation* is an encoded, typically intermediate form of an input. For instance, an embedding vector or vector in the latent space is a representation of the input, as previously discussed. However, representations can also be produced by simpler procedures. For example, one-hot encoded vectors are considered representations of an input.

The key idea is that the representation captures some essential features or characteristics of the original data to make it useful for further analysis or processing.

Exercises

1-1. Suppose we're training a convolutional network with five convolutional layers followed by three fully connected (FC) layers, similar to AlexNet (*https://en.wikipedia.org/wiki/AlexNet*), as illustrated in Figure 1-4.

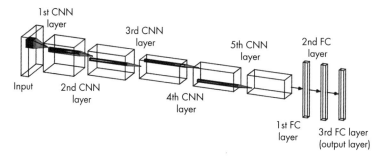

Figure 1-4: An illustration of AlexNet

We can think of these fully connected layers as two hidden layers and an output layer in a multilayer perceptron. Which of the neural network layers can be utilized to produce useful embeddings? Interested readers can find more details about the AlexNet architecture and implementation in the original publication by Alex Krizhevsky, Ilya Sutskever, and Geoffrey Hinton.

1-2. Name some types of input representations that are not embeddings.

References

- The original paper describing the AlexNet architecture and implementation: Alex Krizhevsky, Ilya Sutskever, and Geoffrey Hinton, "ImageNet Classification with Deep Convolutional Neural Networks" (2012), *https://papers.nips.cc/paper/4824-imagenet-classification-with -deep-convolutional-neural-networks*.

2

SELF-SUPERVISED LEARNING

What is self-supervised learning, when is it useful, and what are the main approaches to implementing it?

Self-supervised learning is a pretraining procedure that lets neural networks leverage large, unlabeled datasets in a supervised fashion. This chapter compares self-supervised learning to transfer learning, a related method for pretraining neural networks, and discusses the practical applications of self-supervised learning. Finally, it outlines the main categories of self-supervised learning.

Self-Supervised Learning vs. Transfer Learning

Self-supervised learning is related to transfer learning, a technique in which a model pretrained on one task is reused as the starting point for a model on a second task. For example, suppose we are interested in training an image classifier to classify bird species. In transfer learning, we would pretrain a convolutional neural network on the ImageNet dataset, a large, labeled image dataset with many different categories, including various objects and animals. After pretraining on the general ImageNet dataset, we would take that pretrained model and train it on the smaller, more specific target dataset that contains the bird species of interest. (Often, we just have to change the class-specific output layer, but we can otherwise adopt the pretrained network as is.)

Figure 2-1 illustrates the process of transfer learning.

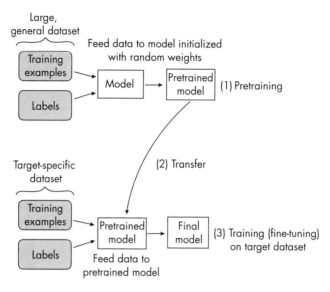

Figure 2-1: Pretraining with conventional transfer learning

Self-supervised learning is an alternative approach to transfer learning in which the model is pretrained not on labeled data but on *unlabeled* data. We consider an unlabeled dataset for which we do not have label information, and then we find a way to obtain labels from the dataset's structure to formulate a prediction task for the neural network, as illustrated in Figure 2-2. These self-supervised training tasks are also called *pretext tasks*.

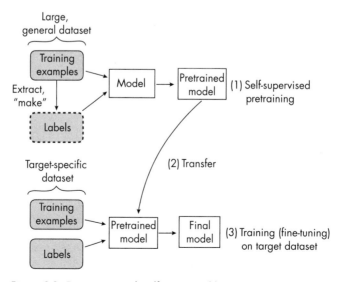

Figure 2-2: Pretraining with self-supervised learning

The main difference between transfer learning and self-supervised learning lies in how we obtain the labels during step 1 in Figures 2-1 and 2-2. In transfer learning, we assume that the labels are provided along with the dataset; they are typically created by human labelers. In self-supervised learning, the labels can be directly derived from the training examples.

A self-supervised learning task could be a missing-word prediction in a natural language processing context. For example, given the sentence "It is beautiful and sunny outside," we can mask out the word *sunny*, feed the network the input "It is beautiful and [MASK] outside," and have the network predict the missing word in the "[MASK]" location. Similarly, we could remove image patches in a computer vision context and have the neural network fill in the blanks. These are just two examples of self-supervised learning tasks; many more methods and paradigms for this type of learning exist.

In sum, we can think of self-supervised learning on the pretext task as *representation learning*. We can take the pretrained model to fine-tune it on the target task (also known as the *downstream* task).

Leveraging Unlabeled Data

Large neural network architectures require large amounts of labeled data to perform and generalize well. However, for many problem areas, we don't have access to large labeled datasets. With self-supervised learning, we can leverage unlabeled data. Hence, self-supervised learning is likely to be useful when working with large neural networks and with a limited quantity of labeled training data.

Transformer-based architectures that form the basis of LLMs and vision transformers are known to require self-supervised learning for pretraining to perform well.

For small neural network models such as multilayer perceptrons with two or three layers, self-supervised learning is typically considered neither useful nor necessary.

Self-supervised learning likewise isn't useful in traditional machine learning with nonparametric models such as tree-based random forests or gradient boosting. Conventional tree-based methods do not have a fixed parameter structure (in contrast to the weight matrices, for example). Thus, conventional tree-based methods are not capable of transfer learning and are incompatible with self-supervised learning.

Self-Prediction and Contrastive Self-Supervised Learning

There are two main categories of self-supervised learning: self-prediction and contrastive self-supervised learning. In *self-prediction*, illustrated in Figure 2-3, we typically change or hide parts of the input and train the model to reconstruct the original inputs, such as by using a perturbation mask that obfuscates certain pixels in an image.

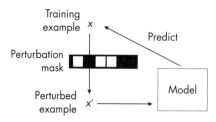

Figure 2-3: Self-prediction after applying a perturbation mask

A classic example is a denoising autoencoder that learns to remove noise from an input image. Alternatively, consider a masked autoencoder that reconstructs the missing parts of an image, as shown in Figure 2-4.

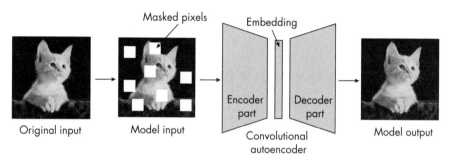

Figure 2-4: A masked autoencoder reconstructing a masked image

Missing (masked) input self-prediction methods are also commonly used in natural language processing contexts. Many generative LLMs, such as GPT, are trained on a next-word prediction pretext task (GPT will be discussed at greater length in Chapters 14 and 17). Here, we feed the network text fragments, where it has to predict the next word in the sequence (as we'll discuss further in Chapter 17).

In *contrastive self-supervised learning*, we train the neural network to learn an embedding space where similar inputs are close to each other and dissimilar inputs are far apart. In other words, we train the network to produce embeddings that minimize the distance between similar training inputs and maximize the distance between dissimilar training examples.

Let's discuss contrastive learning using concrete example inputs. Suppose we have a dataset consisting of random animal images. First, we draw a random image of a cat (the network does not know the label, because we assume that the dataset is unlabeled). We then augment, corrupt, or perturb this cat image, such as by adding a random noise layer and cropping it differently, as shown in Figure 2-5.

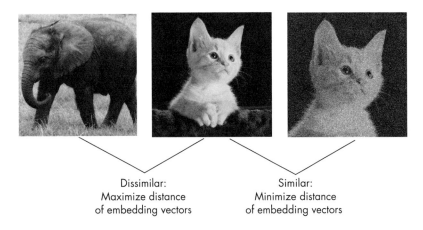

Dissimilar:
Maximize distance
of embedding vectors

Similar:
Minimize distance
of embedding vectors

Figure 2-5: Image pairs encountered in contrastive learning

The perturbed cat image in this figure still shows the same cat, so we want the network to produce a similar embedding vector. We also consider a random image drawn from the training set (for example, an elephant, but again, the network doesn't know the label).

For the cat-elephant pair, we want the network to produce dissimilar embeddings. This way, we implicitly force the network to capture the image's core content while being somewhat agnostic to small differences and noise. For example, the simplest form of a contrastive loss is the L_2-norm (Euclidean distance) between the embeddings produced by model $M(\cdot)$. Let's say we update the model weights to decrease the distance $||M(\text{cat}) - M(\text{cat}')||_2$ and increase the distance $||M(cat) - M(elephant)||_2$.

Figure 2-6 summarizes the central concept behind contrastive learning for the perturbed image scenario. The model is shown twice, which is known as a *siamese network* setup. Essentially, the same model is utilized in two instances: first, to generate the embedding for the original training example, and second, to produce the embedding for the perturbed version of the sample.

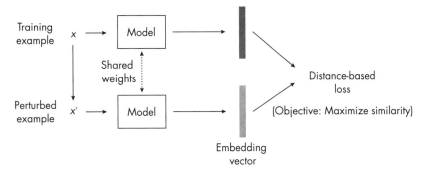

Figure 2-6: Contrastive learning

This example outlines the main idea behind contrastive learning, but many subvariants exist. Broadly, we can categorize these into *sample* contrastive and *dimension* contrastive methods. The elephant-cat example in Figure 2-6 illustrates a sample contrastive method, where we focus on learning embeddings to minimize and maximize distances between training pairs. In *dimension*-contrastive approaches, on the other hand, we focus on making only certain variables in the embedding representations of similar training pairs appear close to each other while maximizing the distance of others.

Exercises

2-1. How could we apply self-supervised learning to video data?

2-2. Can self-supervised learning be used for tabular data represented as rows and columns? If so, how could we approach this?

References

- For more on the ImageNet dataset: *https://en.wikipedia.org/wiki/ImageNet*.

- An example of a contrastive self-supervised learning method: Ting Chen et al., "A Simple Framework for Contrastive Learning of Visual Representations" (2020), *https://arxiv.org/abs/2002.05709*.

- An example of a dimension-contrastive method: Adrien Bardes, Jean Ponce, and Yann LeCun, "VICRegL: Self-Supervised Learning of Local Visual Features" (2022), *https://arxiv.org/abs/2210.01571*.

- If you plan to employ self-supervised learning in practice: Randall Balestriero et al., "A Cookbook of Self-Supervised Learning" (2023), *https://arxiv.org/abs/2304.12210*.

- A paper proposing a method of transfer learning and self-supervised learning for relatively small multilayer perceptrons on tabular datasets: Dara Bahri et al., "SCARF: Self-Supervised Contrastive Learning Using Random Feature Corruption" (2021), *https://arxiv.org/abs/2106.15147*.

- A second paper proposing such a method: Roman Levin et al., "Transfer Learning with Deep Tabular Models" (2022), *https://arxiv.org/abs/2206.15306*.

3

FEW-SHOT LEARNING

What is few-shot learning? How does it differ from the conventional training procedure for supervised learning?

Few-shot learning is a type of supervised learning for small training sets with a very small example-to-class ratio. In regular supervised learning, we train models by iterating over a training set where the model always sees a fixed set of classes. In few-shot learning, we are working on a support set from which we create multiple training tasks to assemble training episodes, where each training task consists of different classes.

Datasets and Terminology

In supervised learning, we fit a model on a training dataset and evaluate it on a test dataset. The training set typically contains a relatively large number of examples per class. For example, in a supervised learning context, the Iris dataset, which has 50 examples per class, is considered a tiny dataset. For deep learning models, on the other hand, even a dataset like MNIST that has 5,000 training examples per class is considered very small.

In few-shot learning, the number of examples per class is much smaller. When specifying the few-shot learning task, we typically use the term N-*way* K-*shot*, where N stands for the number of classes and K stands for the number of examples per class. The most common values are $K = 1$ or $K = 5$. For instance, in a 5-way 1-shot problem, there are five classes with only one example each. Figure 3-1 depicts a 3-way 1-shot setting to illustrate the concept with a smaller example.

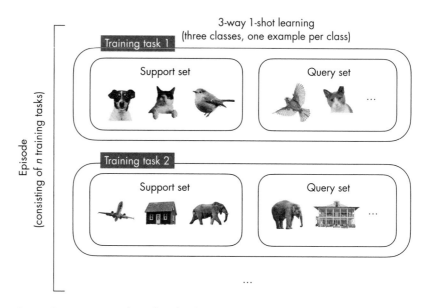

Figure 3-1: Training tasks in few-shot learning

Rather than fitting the model to the training dataset, we can think of few-shot learning as "learning to learn." In contrast to supervised learning, few-shot learning uses not a training dataset but a so-called *support set*, from which we sample training tasks that mimic the use-case scenario during prediction. With each training task comes a query image to be classified. The model is trained on several training tasks from the support set; this is called an *episode*.

Next, during testing, the model receives a new task with classes different from those seen during training. The classes encountered in training are also called *base classes*, and the support set during training is also often called the *base set*. Again, the task is to classify the query images. Test tasks are similar to training tasks, except that none of the classes during testing overlap with those encountered during training, as illustrated in Figure 3-2.

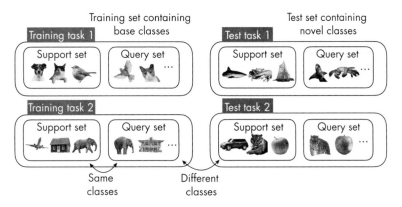

Figure 3-2: Classes seen during training and testing

As Figure 3-2 shows, the support and query sets contain different images from the same class during training. The same is true during testing. However, notice that the classes in the support and query sets differ from the support and query sets encountered during training.

There are many different types of few-shot learning. In the most common, *meta-learning*, training is essentially about updating the model's parameters such that it can *adapt* well to a new task. On a high level, one few-shot learning strategy is to learn a model that produces embeddings where we can find the target class via a nearest-neighbor search among the images in the support set. Figure 3-3 illustrates this approach.

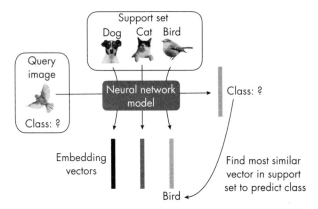

Figure 3-3: Learning embeddings that are suitable for classification

The model learns how to produce good embeddings from the support set to classify the query image based on finding the most similar embedding vector.

Exercises

3-1. MNIST (*https://en.wikipedia.org/wiki/MNIST_database*) is a classic and popular machine learning dataset consisting of 50,000 handwritten digits from 10 classes corresponding to the digits 0 to 9. How can we partition the MNIST dataset for a one-shot classification context?

3-2. What are some real-world applications or use cases for few-shot learning?

4

THE LOTTERY TICKET HYPOTHESIS

What is the lottery ticket hypothesis, and, if it holds true, how is it useful in practice?

The lottery ticket hypothesis is a concept in neural network training that posits that within a randomly initialized neural network, there exists a subnetwork (or "winning ticket") that can, when trained separately, achieve the same accuracy on a test set as the full network after being trained for the same number of steps. This idea was first proposed by Jonathan Frankle and Michael Carbin in 2018.

This chapter illustrates the lottery hypothesis step by step, then goes over *weight pruning*, one of the key techniques to create smaller subnetworks as part of the lottery hypothesis methodology. Lastly, it discusses the practical implications and limitations of the hypothesis.

The Lottery Ticket Training Procedure

Figure 4-1 illustrates the training procedure for the lottery ticket hypothesis in four steps, which we'll discuss one by one to help clarify the concept.

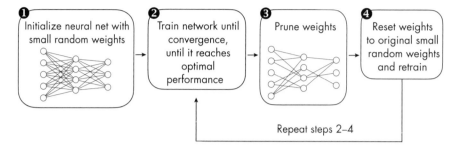

Figure 4-1: The lottery hypothesis training procedure

In Figure 4-1, we start with a large neural network ❶ that we train until convergence ❷, meaning we put in our best efforts to make it perform as well as possible on a target dataset (for example, minimizing training loss and maximizing classification accuracy). This large neural network is initialized as usual using small random weights.

Next, as shown in Figure 4-1, we prune the neural network's weight parameters ❸, removing them from the network. We can do this by setting the weights to zero to create sparse weight matrices. Here, we can either prune individual weights, known as *unstructured* pruning, or prune larger "chunks" from the network, such as entire convolutional filter channels. This is known as *structured* pruning.

The original lottery hypothesis approach follows a concept known as *iterative magnitude pruning*, where the weights with the lowest magnitudes are removed in an iterative fashion. (We will revisit this concept in Chapter 6 when discussing techniques to reduce overfitting.)

After the pruning step, we reset the weights to the original small random values used in step 1 in Figure 4-1 and train the pruned network ❹. It's worth emphasizing that we do not reinitialize the pruned network with any small random weights (as is typical for iterative magnitude pruning), and instead we reuse the weights from step 1.

We then repeat the pruning steps 2 through 4 until we reach the desired network size. For example, in the original lottery ticket hypothesis paper, the authors successfully reduced the network to 10 percent of its original size without sacrificing classification accuracy. As a nice bonus, the pruned (sparse) network, referred to as the *winning ticket*, even demonstrated improved generalization performance compared to the original (large and dense) network.

Practical Implications and Limitations

If it's possible to identify smaller subnetworks that have the same predictive performance as their up-to-10-times-larger counterparts, this can have significant implications for both neural training and inference. Given the ever-growing size of modern neural network architectures, this can help cut training costs and infrastructure.

Sound too good to be true? Maybe. If winning tickets can be identified efficiently, this would be very useful in practice. However, at the time of writing, there is no way to find the winning tickets without training the original network. Including the pruning steps would make this even more expensive than a regular training procedure. Moreover, after the publication of the original paper, researchers found that the original weight initialization may not work to find winning tickets for larger-scale networks, and additional experimentation with the initial weights of the pruned networks is required.

The good news is that winning tickets do exist. Even if it's currently not possible to identify them without training their larger neural network counterparts, they can be used for more efficient inference after training.

Exercises

4-1. Suppose we're trying out the lottery ticket hypothesis approach and find that the performance of the subnetwork is not very good (compared to the original network). What next steps might we try?

4-2. The simplicity and efficiency of the rectified linear unit (ReLU) activation function have made it one of the most popular activation functions in neural network training, particularly in deep learning, where it helps to mitigate problems like the vanishing gradient. The ReLU activation function is defined by the mathematical expression $\max(0, x)$. This means that if the input x is positive, the function returns x, but if the input is negative or 0, the function returns 0. How is the lottery ticket hypothesis related to training a neural network with ReLU activation functions?

References

- The paper proposing the lottery ticket hypothesis: Jonathan Frankle and Michael Carbin, "The Lottery Ticket Hypothesis: Finding Sparse, Trainable Neural Networks" (2018), *https://arxiv.org/abs/1803.03635*.

- The paper proposing structured pruning for removing larger parts, such as entire convolutional filters, from a network: Hao Li et al., "Pruning Filters for Efficient ConvNets" (2016), *https://arxiv.org/abs/1608.08710*.

- Follow-up work on the lottery hypothesis, showing that the original weight initialization may not work to find winning tickets for larger-scale networks, and additional experimentation with the initial weights of the pruned networks is required: Jonathan Frankle et al., "Linear Mode Connectivity and the Lottery Ticket Hypothesis" (2019), *https://arxiv.org/abs/1912.05671*.

- An improved lottery ticket hypothesis algorithm that finds smaller networks that match the performance of a larger network exactly: Vivek Ramanujan et al., "What's Hidden in a Randomly Weighted Neural Network?" (2020), *https://arxiv.org/abs/1911.13299*.

5

REDUCING OVERFITTING WITH DATA

Suppose we train a neural network classifier in a supervised fashion and notice that it suffers from overfitting. What are some of the common ways to reduce overfitting in neural networks through the use of altered or additional data?

Overfitting, a common problem in machine learning, occurs when a model fits the training data too closely, learning its noise and outliers rather than the underlying pattern. As a result, the model performs well on the training data but poorly on unseen or test data. While it is ideal to prevent overfitting, it's often not possible to completely eliminate it. Instead, we aim to reduce or minimize overfitting as much as possible.

The most successful techniques for reducing overfitting revolve around collecting more high-quality labeled data. However, if collecting more labeled data is not feasible, we can augment the existing data or leverage unlabeled data for pretraining.

Common Methods

This chapter summarizes the most prominent examples of dataset-related techniques that have stood the test of time, grouping them into the following categories: collecting more data, data augmentation, and pretraining.

Collecting More Data

One of the best ways to reduce overfitting is to collect more (good-quality) data. We can plot learning curves to find out whether a given model would benefit from more data. To construct a learning curve, we train the model to different training set sizes (10 percent, 20 percent, and so on) and evaluate the trained model on the same fixed-size validation or test set. As shown in Figure 5-1, the validation accuracy increases as the training set sizes increase. This indicates that we can improve the model's performance by collecting more data.

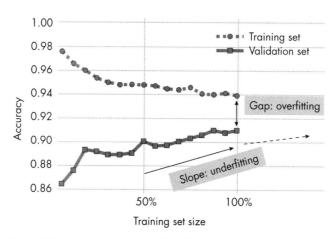

Figure 5-1: The learning curve plot of a model fit to different training set sizes

The gap between training and validation performance indicates the degree of overfitting—the more extensive the gap, the more overfitting occurs. Conversely, the slope indicating an improvement in the validation performance suggests the model is underfitting and can benefit from more data. Typically, additional data can decrease both underfitting and overfitting.

Data Augmentation

Data augmentation refers to generating new data records or features based on existing data. It allows for the expansion of a dataset without additional data collection.

Data augmentation allows us to create different versions of the original input data, which can improve the model's generalization performance. Why? Augmented data can help the model improve its ability to generalize, since it makes it harder to memorize spurious information via training examples or features—or, in the case of image data, exact pixel values for specific pixel locations. Figure 5-2 highlights common image data augmentation techniques, including increasing brightness, flipping, and cropping.

Figure 5-2: A selection of different image data augmentation techniques

Data augmentation is usually standard for image data (see Figure 5-2) and text data (discussed further in Chapter 15), but data augmentation methods for tabular data also exist.

Instead of collecting more data or augmenting existing data, it is also possible to generate new, synthetic data. While more common for image data and text, generating synthetic data is also possible for tabular datasets.

Pretraining

As discussed in Chapter 2, self-supervised learning lets us leverage large, unlabeled datasets to pretrain neural networks. This can also help reduce overfitting on the smaller target datasets.

As an alternative to self-supervised learning, traditional transfer learning on large labeled datasets is also an option. Transfer learning is most effective if the labeled dataset is closely related to the target domain. For instance, if we train a model to classify bird species, we can pretrain a network on a large, general animal classification dataset. However, if such a large animal classification dataset is unavailable, we can also pretrain the model on the relatively broad ImageNet dataset.

A dataset may be extremely small and unsuitable for supervised learning—for example, if it contains only a handful of labeled examples per class. If our classifier needs to operate in a context where the collection of additional labeled data is not feasible, we may also consider few-shot learning.

Other Methods

The previous sections covered the main approaches to using and modifying datasets to reduce overfitting. However, this is not an exhaustive list. Other common techniques include:

- Feature engineering and normalization
- The inclusion of adversarial examples and label or feature noise
- Label smoothing
- Smaller batch sizes
- Data augmentation techniques such as Mixup, Cutout, and CutMix

The next chapter covers additional techniques to reduce overfitting from a model perspective, and it concludes by discussing which regularization techniques we should consider in practice.

Exercises

5-1. Suppose we train an XGBoost model to classify images based on manually extracted features obtained from collaborators. The dataset of labeled training examples is relatively small, but fortunately, our collaborators also have a labeled training set from an older project on a related domain. We're considering implementing a transfer learning approach to train the XGBoost model. Is this a feasible option? If so, how could we do it? (Assume we are allowed to use only XGBoost and not another classification algorithm or model.)

5-2. Suppose we're working on the image classification problem of implementing MNIST-based handwritten digit recognition. We've added a decent amount of data augmentation to try to reduce overfitting. Unfortunately, we find that the classification accuracy is much worse than it was before the augmentation. What are some potential reasons for this?

References

- A paper on data augmentation for tabular data: Derek Snow, "DeltaPy: A Framework for Tabular Data Augmentation in Python" (2020), *https://github.com/firmai/deltapy*.
- The paper proposing the GReaT method for generating synthetic tabular data using an auto-regressive generative large language model: Vadim Borisov et al., "Language Models Are Realistic Tabular Data Generators" (2022), *https://arxiv.org/abs/2210.06280*.
- The paper proposing the TabDDPM method for generating synthetic tabular data using a diffusion model: Akim Kotelnikov et al., "TabDDPM: Modelling Tabular Data with Diffusion Models" (2022), *https://arxiv.org/abs/2209.15421*.

- Scikit-learn's user guide offers a section on preprocessing data, featuring techniques like feature scaling and normalization that can enhance your model's performance: *https://scikit-learn.org/stable/modules/preprocessing.html*.

- A survey on methods for robustly training deep models with noisy labels that explores techniques to mitigate the impact of incorrect or misleading target values: Bo Han et al., "A Survey of Label-noise Representation Learning: Past, Present and Future" (2020), *https://arxiv.org/abs/2011.04406*.

- Theoretical and empirical evidence to support the idea that controlling the ratio of batch size to learning rate in stochastic gradient descent is crucial for good modeling performance in deep neural networks: Fengxiang He, Tongliang Liu, and Dacheng Tao, "Control Batch Size and Learning Rate to Generalize Well: Theoretical and Empirical Evidence" (2019), *https://dl.acm.org/doi/abs/10.5555/3454287.3454390*.

- Inclusion of adversarial examples, which are input samples designed to mislead the model, can improve prediction performance by making the model more robust: Cihang Xie et al., "Adversarial Examples Improve Image Recognition" (2019), *https://arxiv.org/abs/1911.09665*.

- Label smoothing is a regularization technique that mitigates the impact of potentially incorrect labels in the dataset by replacing hard 0 and 1 classification targets with softened values: Rafael Müller, Simon Kornblith, and Geoffrey Hinton, "When Does Label Smoothing Help?" (2019), *https://arxiv.org/abs/1906.02629*.

- Mixup, a popular method that trains neural networks on blended data pairs to improve generalization and robustness: Hongyi Zhang et al., "Mixup: Beyond Empirical Risk Minimization" (2018), *https://arxiv.org/abs/1710.09412*.

6

REDUCING OVERFITTING WITH MODEL MODIFICATIONS

Suppose we train a neural network classifier in a supervised fashion and already employ various dataset-related techniques to mitigate overfitting. How can we change the model or make modifications to the training loop to further reduce the effect of overfitting?

The most successful approaches against overfitting include regularization techniques like dropout and weight decay. As a rule of thumb, models with a larger number of parameters require more training data to generalize well. Hence, decreasing the model size and capacity can sometimes also help reduce overfitting. Lastly, building ensemble models is among the most effective ways to combat overfitting, but it comes with increased computational expense.

This chapter outlines the key ideas and techniques for several categories of reducing overfitting with model modifications and then compares them to one another. It concludes by discussing how to choose between all types of overfitting reduction methods, including those discussed in the previous chapter.

Common Methods

The various model- and training-related techniques to reduce overfitting can be grouped into three broad categories: (1) adding regularization, (2) choosing smaller models, and (3) building ensemble models.

Regularization

We can interpret regularization as a penalty against complexity. Classic regularization techniques for neural networks include L_2 regularization and the related weight decay method. We implement L_2 regularization by adding a penalty term to the loss function that is minimized during training. This added term represents the size of the weights, such as the squared sum of the weights. The following formula shows an L_2 regularized loss

$$RegularizedLoss = Loss + \frac{\lambda}{n} \sum_j w_j^2$$

where λ is a hyperparameter that controls the regularization strength.

During backpropagation, the optimizer minimizes the modified loss, now including the additional penalty term, which leads to smaller model weights and can improve generalization to unseen data.

Weight decay is similar to L_2 regularization but is applied to the optimizer directly rather than modifying the loss function. Since weight decay has the same effect as L_2 regularization, the two methods are often used synonymously, but there may be subtle differences depending on the implementation details and optimizer.

Many other techniques have regularizing effects. For brevity's sake, we'll discuss just two more widely used methods: dropout and early stopping.

Dropout reduces overfitting by randomly setting some of the activations of the hidden units to zero during training. Consequently, the neural network cannot rely on particular neurons to be activated. Instead, it learns to use a larger number of neurons and multiple independent representations of the same data, which helps to reduce overfitting.

In early stopping, we monitor the model's performance on a validation set during training and stop the training process when the performance on the validation set begins to decline, as illustrated in Figure 6-1.

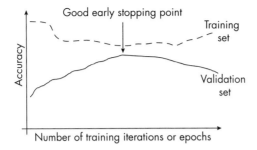

Figure 6-1: Early stopping

In Figure 6-1, we can see that the validation accuracy increases as the training and validation accuracy gap closes. The point where the training and validation accuracy is closest is the point with the least amount of overfitting, which is usually a good point for early stopping.

Smaller Models

Classic bias-variance theory suggests that reducing model size can reduce overfitting. The intuition behind this theory is that, as a general rule of thumb, the smaller the number of model parameters, the smaller its capacity to memorize or overfit to noise in the data. The following paragraphs discuss methods to reduce the model size, including pruning, which removes parameters from a model, and knowledge distillation, which transfers knowledge to a smaller model.

Besides reducing the number of layers and shrinking the layers' widths as a hyperparameter tuning procedure, another approach to obtaining smaller models is *iterative pruning*, in which we train a large model to achieve good performance on the original dataset. We then iteratively remove parameters of the model, retraining it on the dataset such that it maintains the same predictive performance as the original model. (The lottery ticket hypothesis, discussed in Chapter 4, uses iterative pruning.)

Another common approach to obtaining smaller models is *knowledge distillation*. The general idea behind this approach is to transfer knowledge from a large, more complex model (the *teacher*) to a smaller model (the *student*). Ideally, the student achieves the same predictive accuracy as the teacher, but it does so more efficiently due to the smaller size. As a nice side effect, the smaller student may overfit less than the larger teacher model.

Figure 6-2 diagrams the original knowledge distillation process. Here, the teacher is first trained in a regular supervised fashion to classify the examples in the dataset well, using a conventional cross-entropy loss between the predicted scores and ground truth class labels. While the smaller student network is trained on the same dataset, the training objective is to minimize both (a) the cross entropy between the outputs and the class labels and (b) the difference between its outputs and the teacher outputs (measured using *Kullback–Leibler* divergence, which quantifies the difference between two probability distributions by calculating how much one distribution diverges from the other in terms of information content).

(1) Train teacher model to achieve high classification accuracy

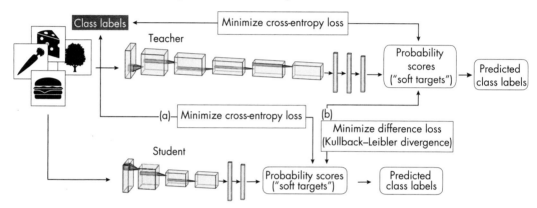

(2) Train student model on same dataset to minimize combined loss
consisting of (a) cross-entropy and (b) Kullback–Leibler divergence

Figure 6-2: The original knowledge distillation process

By minimizing the Kullback–Leibler divergence—the difference between the teacher and student score distributions—the student learns to mimic the teacher while being smaller and more efficient.

Caveats with Smaller Models

While pruning and knowledge distillation can also enhance a model's generalization performance, these techniques are not primary or effective ways of reducing overfitting.

Early research results indicate that pruning and knowledge distillation can improve the generalization performance, presumably due to smaller model sizes. However, counterintuitively, recent research studying phenomena like double descent and grokking also showed that larger, overparameterized models have improved generalization performance if they are trained beyond the point of overfitting. *Double descent* refers to the phenomenon in which models with either a small or an extremely large number of parameters have good generalization performance, while models with a number of parameters equal to the number of training data points have poor generalization performance. *Grokking* reveals that as the size of a dataset decreases, the need for optimization increases, and generalization performance can improve well past the point of overfitting.

How can we reconcile the observation that pruned models can exhibit better generalization performance with contradictory observations from studies of double descent and grokking? Researchers recently showed that the improved training process partly explains the reduction of overfitting due to pruning. Pruning involves more extended training periods and a replay of learning rate schedules that may be partly responsible for the improved generalization performance.

Pruning and knowledge distillation remain excellent ways to improve the computational efficiency of a model. However, while they can also enhance a model's generalization performance, these techniques are not primary or effective ways of reducing overfitting.

Ensemble Methods

Ensemble methods combine predictions from multiple models to improve the overall prediction performance. However, the downside of using multiple models is an increased computational cost.

We can think of ensemble methods as asking a committee of experts to weigh in on a decision and then combining their judgments in some way to make a final decision. Members in a committee often have different backgrounds and experiences. While they tend to agree on basic decisions, they can overrule bad decisions by majority rule. This doesn't mean that the majority of experts is always right, but there is a good chance that the majority of the committee is more often right, on average, than every single member.

The most basic example of an ensemble method is majority voting. Here, we train k different classifiers and collect the predicted class label from each of these k models for a given input. We then return the most frequent class label as the final prediction. (Ties are usually resolved using a confidence score, randomly picking a label, or picking the class label with the lowest index.)

Ensemble methods are more prevalent in classical machine learning than deep learning because it is more computationally expensive to employ multiple models than to rely on a single one. In other words, deep neural networks require significant computational resources, making them less suitable for ensemble methods.

Random forests and gradient boosting are popular examples of ensemble methods. However, by using majority voting or stacking, for example, we can combine any group of models: an ensemble may consist of a support vector machine, a multilayer perceptron, and a nearest-neighbor classifier. Here, stacking (also known as *stacked generalization*) is a more advanced variant of majority voting that involves training a new model to combine the predictions of several other models rather than obtaining the label by majority vote.

A popular industry technique is to build models from *k-fold cross-validation*, a model evaluation technique in which we train and evaluate a model on k training folds. We then compute the average performance metric across all k iterations to estimate the overall performance measure of the model. After evaluation, we can either train the model on the entire training dataset or combine the individual models as an ensemble, as shown in Figure 6-3.

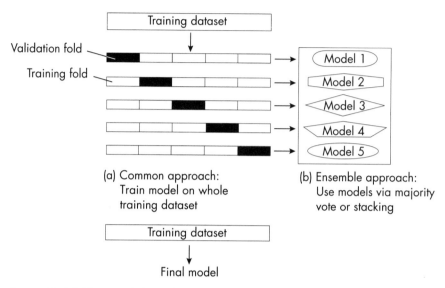

Figure 6-3: k-fold cross-validation for creating model ensembles

As shown in Figure 6-3, the k-fold ensemble approach trains each of the k models on the respective $k - 1$ training folds in each round. After evaluating the models on the validation folds, we can combine them into a majority vote classifier or build an ensemble using stacking, a technique that combines multiple classification or regression models via a meta-model.

While the ensemble approach can potentially reduce overfitting and improve robustness, this approach is not always suitable. For instance, potential downsides include managing and deploying an ensemble of models, which can be more complex and computationally expensive than using a single model.

Other Methods

So far, this book has covered some of the most prominent techniques to reduce overfitting. Chapter 5 covered techniques that aim to reduce overfitting from a data perspective. Additional techniques for reducing overfitting with model modifications include skip-connections (found in residual networks, for example), look-ahead optimizers, stochastic weight averaging, multitask learning, and snapshot ensembles.

While they are not originally designed to reduce overfitting, layer input normalization techniques such as batch normalization (BatchNorm) and layer normalization (LayerNorm) can stabilize training and often have a regularizing effect that reduces overfitting. Weight normalization, which normalizes the model weights instead of layer inputs, could also lead to better generalization performance. However, this effect is less direct since weight normalization (WeightNorm) doesn't explicitly act as a regularizer like weight decay does.

Choosing a Regularization Technique

Improving data quality is an essential first step in reducing overfitting. However, for recent deep neural networks with large numbers of parameters, we need to do more to achieve an acceptable level of overfitting. Therefore, data augmentation and pretraining, along with established techniques such as dropout and weight decay, remain crucial overfitting reduction methods.

In practice, we can and should use multiple methods at once to reduce overfitting for an additive effect. To achieve the best results, treat selecting these techniques as a hyperparameter optimization problem.

Exercises

6-1. Suppose we're using early stopping as a mechanism to reduce overfitting—in particular, a modern early-stopping variant that creates checkpoints of the best model (for instance, the model with the highest validation accuracy) during training so that we can load it after the training has completed. This mechanism can be enabled in most modern deep learning frameworks. However, a colleague recommends tuning the number of training epochs instead. What are some of the advantages and disadvantages of each approach?

6-2. Ensemble models have been established as a reliable and successful method for decreasing overfitting and enhancing the reliability of predictive modeling efforts. However, there's always a trade-off. What are some of the drawbacks associated with ensemble techniques?

References

- For more on the distinction between L_2 regularization and weight decay: Guodong Zhang et al., "Three Mechanisms of Weight Decay Regularization" (2018), *https://arxiv.org/abs/1810.12281*.

- Research results indicate that pruning and knowledge distillation can improve generalization performance, presumably due to smaller model sizes: Geoffrey Hinton, Oriol Vinyals, and Jeff Dean, "Distilling the Knowledge in a Neural Network" (2015), *https://arxiv.org/abs/1503.02531*.

- Classic bias-variance theory suggests that reducing model size can reduce overfitting: Jerome H. Friedman, Robert Tibshirani, and Trevor Hastie, "Model Selection and Bias-Variance Tradeoff," Chapter 2.9, in *The Elements of Statistical Learning* (Springer, 2009).

- The lottery ticket hypothesis applies knowledge distillation to find smaller networks with the same predictive performance as the original one: Jonathan Frankle and Michael Carbin, "The Lottery Ticket Hypothesis: Finding Sparse, Trainable Neural Networks" (2018), *https://arxiv.org/abs/1803.03635*.

- For more on double descent: *https://en.wikipedia.org/wiki/Double_descent*.

- The phenomenon of grokking indicates that generalization performance can improve well past the point of overfitting: Alethea Power et al., "Grokking: Generalization Beyond Overfitting on Small Algorithmic Datasets" (2022), *https://arxiv.org/abs/2201.02177*.

- Recent research shows that the improved training process partly explains the reduction of overfitting due to pruning: Tian Jin et al., "Pruning's Effect on Generalization Through the Lens of Training and Regularization" (2022), *https://arxiv.org/abs/2210.13738*.

- Dropout was previously discussed as a regularization technique, but it can also be considered an ensemble method that approximates a weighted geometric mean of multiple networks: Pierre Baldi and Peter J. Sadowski, "Understanding Dropout" (2013), *https://proceedings.neurips.cc/paper/2013/hash/71f6278d140af599 e06ad9bf1ba03cb0-Abstract.html*.

- Regularization cocktails need to be tuned on a per-dataset basis: Arlind Kadra et al., "Well-Tuned Simple Nets Excel on Tabular Datasets" (2021), *https://arxiv.org/abs/2106.11189*.

7

MULTI-GPU TRAINING PARADIGMS

What are the different multi-GPU training paradigms, and what are their respective advantages and disadvantages?

Multi-GPU training paradigms can be categorized into two groups: dividing data for parallel processing with multiple GPUs and dividing the model among multiple GPUs to handle memory constraints when the model size surpasses that of a single GPU. Data parallelism falls into the first category, while model parallelism and tensor parallelism fall into the second category. Techniques like pipeline parallelism borrow ideas from both categories. In addition, current software implementations such as DeepSpeed, Colossal AI, and others blend multiple approaches into a hybrid technique.

This chapter introduces several training paradigms and provides advice on which to use in practice.

NOTE *This chapter primarily uses the term* GPUs *to describe the hardware utilized for parallel processing. However, the same concepts and techniques discussed can be applied to other specialized hardware devices, such as tensor processing units (TPUs) or other accelerators, depending on the specific architecture and requirements of the system.*

The Training Paradigms

The following sections discuss the model parallelism, data parallelism, tensor parallelism, and sequence parallelism multi-GPU training paradigms.

Model Parallelism

Model parallelism, or *inter-op parallelism*, is a technique in which different sections of a large model are placed on different GPUs and are computed sequentially, with intermediate results passed between the devices. This allows for the training and execution of models that might not fit entirely on a single device, but it can require intricate coordination to manage the dependencies between different parts of the model.

Model parallelism is perhaps the most intuitive form of parallelization across devices. For example, for a simple neural network that consists of only two layers—a hidden layer and an output layer—we can keep one layer on one GPU and the other layer on another GPU. Of course, this can scale to an arbitrary number of layers and GPUs.

This is a good strategy for dealing with limited GPU memory where the complete network does not fit into one GPU. However, there are more efficient ways of using multiple GPUs, such as tensor parallelism, because the chain-like structure (layer 1 on GPU 1 → layer 2 on GPU 2 → ...) in model parallelism introduces a bottleneck. In other words, a major disadvantage of model parallelism is that the GPUs have to wait for each other. They cannot efficiently work in parallel, as they depend on one other's outputs.

Data Parallelism

Data parallelism has been the default mode for multi-GPU training for several years. Here, we divide a minibatch into smaller microbatches. Each GPU then processes a microbatch separately to compute the loss and loss gradients for the model weights. After the individual devices process the microbatches, the gradients are combined to compute the weight update for the next round.

An advantage of data parallelism over model parallelism is that the GPUs can run in parallel. Each GPU processes a portion of the training minibatch, that is, a microbatch. However, a caveat is that each GPU requires a full copy of the model. This is obviously not feasible if we have large models that don't fit into the GPU's VRAM.

Tensor Parallelism

Tensor parallelism, or *intra-op parallelism*, is a more efficient form of model parallelism. Here, the weight and activation matrices are spread across the devices instead of distributing whole layers across devices: the individual matrices are split, so we split an individual matrix multiplication across GPUs.

We can implement tensor parallelism using basic principles of linear algebra; we can split a matrix multiplication across two GPUs in a row- or column-wise fashion, as illustrated in Figure 7-1 for two GPUs. (This concept can be extended to an arbitrary number of GPUs.)

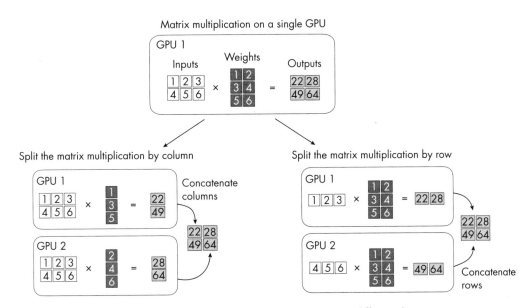

Figure 7-1: Tensor parallelism for distributing matrix multiplication across different devices

Like model parallelism, tensor parallelism allows us to work around memory limitations. At the same time, it also lets us execute operations in parallel, similar to data parallelism.

A small weakness of tensor parallelism is that it can result in high communication overhead between the multiple GPUs across which the matrices are split or sharded. For instance, tensor parallelism requires frequent synchronization of the model parameters across devices, which can slow down the overall training process.

Figure 7-2 compares model, data, and tensor parallelism.

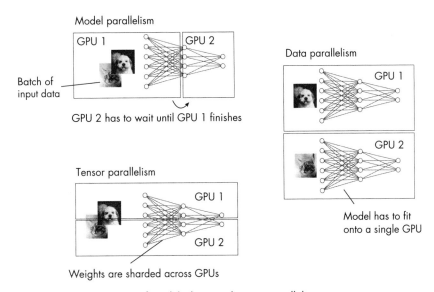

Figure 7-2: A comparison of model, data, and tensor parallelism

In model parallelism, we put different layers onto different GPUs to work around GPU memory limitations. In data parallelism, we split a batch across GPUs to train copies of the model in parallel, averaging gradients for the weight update afterward. In tensor parallelism, we split matrices (inputs and weights) across different GPUs for parallel processing when models are too large to fit into GPU memory.

Pipeline Parallelism

In *pipeline parallelism*, activations are passed during the forward pass, as in model parallelism. The twist is that the gradients of the input tensor are passed backward to prevent the devices from being idle. In a sense, pipeline parallelism is a sophisticated hybrid version of data and model parallelism.

We can think of pipeline parallelism as a form of model parallelism that tries to minimize the sequential computation bottleneck, enhancing the parallelism between the individual layers sitting on different devices. However, pipeline parallelism also borrows ideas from data parallelism, such as splitting minibatches further into microbatches.

Pipeline parallelism is definitely an improvement over model parallelism, though it is not perfect and there will be idle bubbles. A further disadvantage of pipeline parallelism is that it may require significant effort to design and implement the pipeline stages and associated communication patterns. Additionally, the performance gains it generates may not be as substantial as those from other parallelization techniques, such as pure data parallelism, especially for small models or in cases where the communication overhead is high.

For modern architectures that are too large to fit into GPU memory, it is more common nowadays to use a blend of data parallelism and tensor parallelism techniques instead of pipeline parallelism.

Sequence Parallelism

Sequence parallelism aims to address computational bottlenecks when working with long sequences using transformer-based LLMs. More specifically, one shortcoming of transformers is that the self-attention mechanism (the original scaled-dot product attention) scales quadratically with the input sequence length. There are, of course, more efficient alternatives to the original attention mechanism that scale linearly.

However, these efficient self-attention mechanisms are less popular, and most people still prefer the original scaled-dot product attention mechanism as of this writing. Sequence parallelism, illustrated in Figure 7-3, splits the input sequence into smaller chunks to be distributed across GPUs, which aims to reduce computation memory constraints of self-attention mechanisms.

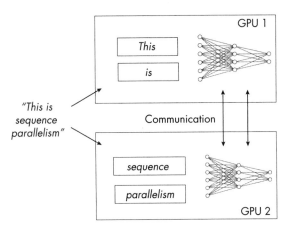

Figure 7-3: Sequence parallelism divides long inputs among GPUs.

How does sequence parallelism relate to the multi-GPU techniques discussed earlier? Sequence parallelism deals specifically with sequential data, tensor parallelism deals with the model's internal structure, and data parallelism deals with how the training data is divided. Theoretically, since each of these parallelism strategies addresses a different aspect of the computational challenge, they can thus be combined in various ways to optimize the training or inference process. Sequence parallelism is not as well studied as other parallelization techniques, however.

While sequence parallelism appears useful in practice, it also introduces additional communication overheads similar to the aforementioned parallelism techniques. Like data parallelism, it requires us to duplicate the model and make sure it fits into the device memory. Another of its disadvantages (depending on the implementation) for multi-GPU training of transformers is that breaking up the input sequence into smaller subsequences can decrease the model's accuracy (mainly when the model is applied to longer sequences).

Recommendations

Practical recommendations depend on the context. If we train small models that fit onto a single GPU, then data parallelism strategies may be the most efficient. Performance gains from pipeline parallelism may not be as significant as those from other parallelization techniques, such as data parallelism, especially for small models or in cases where the communication overhead is high.

If models are too large to fit into the memory of a single GPU, we need to explore model or tensor parallelism. Tensor parallelism is naturally more efficient; the GPUs can work in parallel since there is no sequential dependency as in model parallelism.

Modern multi-GPU strategies also typically combine data parallelism and tensor parallelism.

Exercises

7-1. Suppose we are implementing our own version of tensor parallelism, which works great when we train our model with a standard stochastic gradient descent optimizer. However, when we try the Adam optimizer by Diederik P. Kingma and Jimmy Ba, we encounter an out-of-memory device. What problem might explain this issue?

7-2. Suppose we don't have access to a GPU and are considering using data parallelism on the CPU. Is this a good idea?

References

- The original paper on the Adam optimizer: Diederik P. Kingma and Jimmy Ba, "Adam: A Method for Stochastic Optimization" (2014), *https://arxiv.org/abs/1412.6980*.

- For more on DeepSpeed and Colossal-AI for multi-GPU training: *https://github.com/microsoft/DeepSpeed* and *https://github.com/hpcaitech/ColossalAI*.

- Pipeline parallelism tutorials and research by the DeepSpeed team: *https://www.deepspeed.ai/tutorials/pipeline* and Yanping Huang et al., "GPipe: Efficient Training of Giant Neural Networks Using Pipeline Parallelism" (2018), *https://arxiv.org/abs/1811.06965*.

- The paper proposing sequence parallelism for transformer-based language models: Shenggui Li et al., "Sequence Parallelism: Long Sequence Training from [a] System[s] Perspective" (2022), *https://arxiv.org/abs/2105.13120*.

- The scaled-dot product attention mechanism was proposed with the original transformer architecture: Ashish Vaswani et al., "Attention Is All You Need" (2017), *https://arxiv.org/abs/1706.03762*.

- A survey covering alternatives to the original self-attention mechanism that scale linearly: Yi Tay et al., "Efficient Transformers: A Survey" (2020), *https://arxiv.org/abs/2009.06732*.

- A survey covering additional techniques to improve the training efficiency of transformers: Bohan Zhuang et al., "A Survey on Efficient Training of Transformers" (2023), *https://arxiv.org/abs/2302.01107*.

- Modern multi-GPU strategies typically combine data parallelism and tensor parallelism. Popular examples include DeepSpeed stages 2 and 3, described in this tutorial on the zero redundancy optimizer: *https://www.deepspeed.ai/tutorials/zero/*.

8

THE SUCCESS OF TRANSFORMERS

What are the main factors that have contributed to the success of transformers?

In recent years, transformers have emerged as the most successful neural network architecture, particularly for various natural language processing tasks. In fact, transformers are now on the cusp of becoming state of the art for computer vision tasks as well. The success of transformers can be attributed to several key factors, including their attention mechanisms, ability to be parallelized easily, unsupervised pretraining, and high parameter counts.

The Attention Mechanism

The self-attention mechanism found in transformers is one of the key design components that make transformer-based LLMs so successful. However, transformers are not the first architecture to utilize attention mechanisms.

Attention mechanisms were first developed in the context of image recognition back in 2010, before being adopted to aid the translation of long sentences in recurrent neural networks. (Chapter 16 compares the attention mechanisms found in recurrent neural networks and transformers in greater detail.)

The aforementioned attention mechanism is inspired by human vision, focusing on specific parts of an image (foveal glimpses) at a time to process information hierarchically and sequentially. In contrast, the fundamental mechanism underlying transformers is a self-attention mechanism used for sequence-to-sequence tasks, such as machine translation and text generation. It allows each token in a sequence to attend to all other tokens, thus providing context-aware representations of each token.

What makes attention mechanisms so unique and useful? For the following illustration, suppose we are using an encoder network on a fixed-length representation of the input sequence or image—this can be a fully connected, convolutional, or attention-based encoder.

In a transformer, the encoder uses self-attention mechanisms to compute the importance of each input token relative to other tokens in the sequence, allowing the model to focus on relevant parts of the input sequence. Conceptually, attention mechanisms allow the transformers to attend to different parts of a sequence or image. On the surface, this sounds very similar to a fully connected layer where each input element is connected via a weight with the input element in the next layer. In attention mechanisms, the computation of the attention weights involves comparing each input element to all others. The attention weights obtained by this approach are dynamic and input dependent. In contrast, the weights of a convolutional or fully connected layer are fixed after training, as illustrated in Figure 8-1.

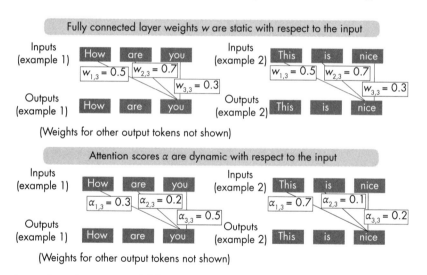

Figure 8-1: The conceptual difference between model weights in fully connected layers (top) and attention scores (bottom)

As the top part of Figure 8-1 shows, once trained, the weights of fully connected layers remain fixed regardless of the input. In contrast, as shown at the bottom, self-attention weights change depending on the inputs, even after a transformer is trained.

Attention mechanisms allow a neural network to selectively weigh the importance of different input features, so the model can focus on the most relevant parts of the input for a given task. This provides a contextual understanding of each word or image token, allowing for more nuanced interpretations, which is one of the aspects that can make transformers work so well.

Pretraining via Self-Supervised Learning

Pretraining transformers via self-supervised learning on large, unlabeled datasets is another key factor in the success of transformers. During pretraining, the transformer model is trained to predict missing words in a sentence or the next sentence in a document, for example. By learning to predict these missing words or the next sentence, the model is forced to learn general representations of language that can be fine-tuned for a wide range of downstream tasks.

While unsupervised pretraining has been highly effective for natural language processing tasks, its effectiveness for computer vision tasks is still an active area of research. (Refer to Chapter 2 for a more detailed discussion of self-supervised learning.)

Large Numbers of Parameters

One noteworthy characteristic of transformers is their large model sizes. For example, the popular 2020 GPT-3 model consists of 175 billion trainable parameters, while other transformers, such as switch transformers, have trillions of parameters.

The scale and number of trainable parameters of transformers are essential factors in their modeling performance, particularly for large-scale natural language processing tasks. For instance, linear scaling laws suggest that the training loss decreases proportionally with an increase in model size, so a doubling of the model size can halve the training loss.

This, in turn, can lead to better performance on the downstream target task. However, it is essential to scale the model size and the number of training tokens equally. This means the number of training tokens should be doubled for every doubling of model size.

Since labeled data is limited, utilizing large amounts of data during unsupervised pretraining is vital.

To summarize, large model sizes and large datasets are critical factors in transformers' success. Additionally, using self-supervised learning, the ability to pretrain transformers is closely tied to using large model sizes and large datasets. This combination has been critical in enabling the success of transformers in a wide range of natural language processing tasks.

Easy Parallelization

Training large models on large datasets requires vast computational resources, and it's key that the computations can be parallelized to utilize these resources.

Fortunately, transformers are easy to parallelize since they take a fixed-length sequence of word or image tokens as input. For instance, the self-attention mechanism used in most transformer architectures involves computing the weighted sum between a pair of input elements. Furthermore, these pair-wise token comparisons can be computed independently, as illustrated in Figure 8-2, making the self-attention mechanism relatively easy to parallelize across different GPU cores.

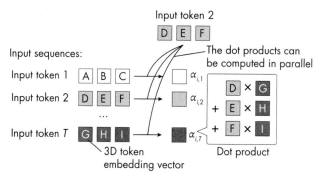

Figure 8-2: A simplified self-attention mechanism without weight parameters

In addition, the individual weight matrices used in the self-attention mechanism (not shown in Figure 8-2) can be distributed across different machines for distributed and parallel computing.

Exercises

8-1. As discussed in this chapter, self-attention is easily parallelizable, yet transformers are considered computationally expensive due to self-attention. How can we explain this contradiction?

8-2. Since self-attention scores represent importance weights for the various input elements, can we consider self-attention to be a form of feature selection?

References

- An example of an attention mechanism in the context of image recognition: Hugo Larochelle and Geoffrey Hinton, "Learning to Combine Foveal Glimpses with a Third-Order Boltzmann Machine" (2010), *https://dl.acm.org/doi/10.5555/2997189.2997328*.

- The paper introducing the self-attention mechanism with the original transformer architecture: Ashish Vaswani et al., "Attention Is All You Need" (2017), *https://arxiv.org/abs/1706.03762*.

- Transformers can have trillions of parameters: William Fedus, Barret Zoph, and Noam Shazeer, "Switch Transformers: Scaling to Trillion Parameter Models with Simple and Efficient Sparsity" (2021), *https://arxiv.org/abs/2101.03961*.

- Linear scaling laws suggest that training loss decreases proportionally with an increase in model size: Jared Kaplan et al., "Scaling Laws for Neural Language Models" (2020), *https://arxiv.org/abs/2001.08361*.

- Research suggests that in transformer-based language models, the training tokens should be doubled for every doubling of model size: Jordan Hoffmann et al., "Training Compute-Optimal Large Language Models" (2022), *https://arxiv.org/abs/2203.15556*.

- For more about the weights used in self-attention and cross-attention mechanisms, check out my blog post: "Understanding and Coding the Self-Attention Mechanism of Large Language Models from Scratch" at *https://sebastianraschka.com/blog/2023/self-attention-from -scratch.html*.

9

GENERATIVE AI MODELS

What are the popular categories of deep generative models in deep learning (also called *generative AI*), and what are their respective downsides?

Many different types of deep generative models have been applied to generating different types of media: images, videos, text, and audio. Beyond these types of media, models can also be repurposed to generate domain-specific data, such as organic molecules and protein structures. This chapter will first define generative modeling and then outline each type of generative model and discuss its strengths and weaknesses.

Generative vs. Discriminative Modeling

In traditional machine learning, there are two primary approaches to modeling the relationship between input data (x) and output labels (y): generative models and discriminative models. *Generative models* aim to capture the underlying probability distribution of the input data $p(x)$ or the joint distribution $p(x, y)$ between inputs and labels. In contrast, *discriminative models* focus on modeling the conditional distribution $p(y|x)$ of the labels given the inputs.

A classic example that highlights the differences between these approaches is to compare the naive Bayes classifier and the logistic regression classifier. Both classifiers estimate the class label probabilities $p(y|x)$ and can be used for classification tasks. However, logistic regression is considered a discriminative model because it directly models the conditional probability distribution $p(y|x)$ of the class labels given the input features without making assumptions about the underlying joint distribution of inputs and labels. Naive Bayes, on the other hand, is considered a generative model because it models the joint probability distribution $p(x, y)$ of the input features x and the output labels y. By learning the joint distribution, a generative model like naive Bayes captures the underlying data generation process, which enables it to generate new samples from the distribution if needed.

Types of Deep Generative Models

When we speak of *deep* generative models or deep generative AI, we often loosen this definition to include all types of models capable of producing realistic-looking data (typically text, images, videos, and sound). The remainder of this chapter briefly discusses the different types of deep generative models used to generate such data.

Energy-Based Models

Energy-based models (EBMs) are a class of generative models that learn an energy function, which assigns a scalar value (energy) to each data point. Lower energy values correspond to more likely data points. The model is trained to minimize the energy of real data points while increasing the energy of generated data points. Examples of EBMs include *deep Boltzmann machines (DBMs)*. One of the early breakthroughs in deep learning, DBMs provide a means to learn complex representations of data. You can think of them as a form of unsupervised pretraining, resulting in models that can then be fine-tuned for various tasks.

Somewhat similar to naive Bayes and logistic regression, DBMs and multilayer perceptrons (MLPs) can be thought of as generative and discriminative counterparts, with DBMs focusing on capturing the data generation process and MLPs focusing on modeling the decision boundary between classes or mapping inputs to outputs.

A DBM consists of multiple layers of hidden nodes, as shown in Figure 9-1. As the figure illustrates, along with the hidden layers, there's usually a visible layer that corresponds to the observable data. This visible layer serves as the input layer where the actual data or features are fed into the network. In addition to using a different learning algorithm than MLPs (contrastive divergence instead of backpropagation), DBMs consist of binary nodes (neurons) instead of continuous ones.

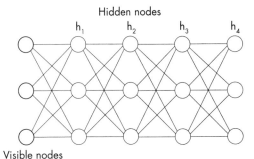

Figure 9-1: A four-layer deep Boltzmann machine with three stacks of hidden nodes

Suppose we are interested in generating images. A DBM can learn the joint probability distribution over the pixel values in a simple image dataset like MNIST. To generate new images, the DBM then samples from this distribution by performing a process called *Gibbs sampling*. Here, the visible layer of the DBM represents the input image. To generate a new image, the DBM starts by initializing the visible layer with random values or, alternatively, uses an existing image as a seed. Then, after completing several Gibbs sampling iterations, the final state of the visible layer represents the generated image.

DBMs played an important historical role as one of the first deep generative models, but they are no longer very popular for generating data. They are expensive and more complicated to train, and they have lower expressivity compared to the newer models described in the following sections, which generally results in lower-quality generated samples.

Variational Autoencoders

Variational autoencoders (VAEs) are built upon the principles of variational inference and autoencoder architectures. *Variational inference* is a method for approximating complex probability distributions by optimizing a simpler, tractable distribution to be as close as possible to the true distribution. *Autoencoders* are unsupervised neural networks that learn to compress input data into a low-dimensional representation (encoding) and subsequently reconstruct the original data from the compressed representation (decoding) by minimizing the reconstruction error.

The VAE model consists of two main submodules: an encoder network and a decoder network. The encoder network takes, for example, an input image and maps it to a latent space by learning a probability distribution over the latent variables. This distribution is typically modeled as a Gaussian with parameters (mean and variance) that are functions of the input image. The decoder network then takes a sample from the learned latent distribution and reconstructs the input image from this sample. The goal of the VAE is to learn a compact and expressive latent representation that captures the essential structure of the input data while being able to generate new images by sampling from the latent space. (See Chapter 1 for more details on latent representations.)

Figure 9-2 illustrates the encoder and decoder submodules of an autoencoder, where x' represents the reconstructed input x. In a standard variational autoencoder, the latent vector is sampled from a distribution that approximates a standard Gaussian distribution.

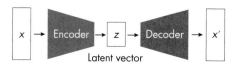

Figure 9-2: An autoencoder

Training a VAE involves optimizing the model's parameters to minimize a loss function composed of two terms: a reconstruction loss and a Kullback–Leibler-divergence (KL-divergence) regularization term. The reconstruction loss ensures that the decoded samples closely resemble the input images, while the KL-divergence term acts as a surrogate loss that encourages the learned latent distribution to be close to a predefined prior distribution (usually a standard Gaussian). To generate new images, we then sample points from the latent space's prior (standard Gaussian) distribution and pass them through the decoder network, which generates new, diverse images that look similar to the training data.

Disadvantages of VAEs include their complicated loss function consisting of separate terms, as well as their often low expressiveness. The latter can result in blurrier images compared to other models, such as generative adversarial networks.

Generative Adversarial Networks

Generative adversarial networks (GANs) are models consisting of interacting subnetworks designed to generate new data samples that are similar to a given set of input data. While both GANs and VAEs are latent variable models that generate data by sampling from a learned latent space, their architectures and learning mechanisms are fundamentally different.

GANs consist of two neural networks, a generator and a discriminator, that are trained simultaneously in an adversarial manner. The generator takes a random noise vector from the latent space as input and generates a synthetic data sample (such as an image). The discriminator's task is to distinguish between real samples from the training data and fake samples generated by the generator, as illustrated in Figure 9-3.

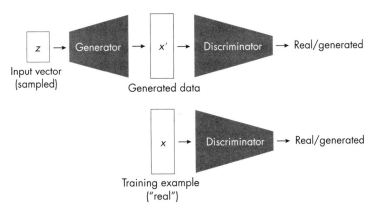

Figure 9-3: A generative adversarial network

The generator in a GAN somewhat resembles the decoder of a VAE in terms of its functionality. During inference, both GAN generators and VAE decoders take random noise vectors sampled from a known distribution (for example, a standard Gaussian) and transform them into synthetic data samples, such as images.

One significant disadvantage of GANs is their unstable training due to the adversarial nature of the loss function and learning process. Balancing the learning rates of the generator and discriminator can be difficult and can often result in oscillations, mode collapse, or non-convergence. The second main disadvantage of GANs is the low diversity of their generated outputs, often due to mode collapse. Here, the generator is able to fool the discriminator successfully with a small set of samples, which are representative of only a small subset of the original training data.

Flow-Based Models

The core concept of *flow-based models*, also known as *normalizing flows*, is inspired by long-standing methods in statistics. The primary goal is to transform a simple probability distribution (like a Gaussian) into a more complex one using invertible transformations.

Although the concept of normalizing flows has been a part of the statistics field for a long time, the implementation of early flow-based deep learning models, particularly for image generation, is a relatively recent development. One of the pioneering models in this area was the *non-linear independent components estimation (NICE)* approach. NICE begins with a simple probability distribution, often something straightforward like a normal distribution. You can think of this as a kind of "random noise," or data with no particular shape or structure. NICE then applies a series of transformations to this simple distribution. Each transformation is designed to make the data look more like the final target (for instance, the distribution of real-world images). These transformations are "invertible," meaning we can always reverse them back to the original simple distribution. After several successive transformations, the simple distribution has morphed into a complex distribution that closely matches the distribution of the target data (such as images). We can now generate new data that looks like the target data by picking random points from this complex distribution.

Figure 9-4 illustrates the concept of a flow-based model, which maps the complex input distribution to a simpler distribution and back.

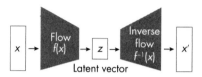

Figure 9-4: A flow-based model

At first glance, the illustration is very similar to the VAE illustration in Figure 9-2. However, while VAEs use neural network encoders like convolutional neural networks, the flow-based model uses simpler decoupling layers, such as simple linear transformations. Additionally, while the decoder in a VAE is independent of the encoder, the data-transforming functions in the flow-based model are mathematically inverted to obtain the outputs.

Unlike VAEs and GANs, flow-based models provide exact likelihoods, which gives us insights into how well the generated samples fit the training data distribution. This can be useful in anomaly detection or density estimation, for example. However, the quality of flow-based models for generating image data is usually lower than GANs. Flow-based models also often require more memory and computational resources than GANs or VAEs since they must store and compute inverses of transformations.

Autoregressive Models

Autoregressive models are designed to predict the next value based on current (and past) values. LLMs for text generation, like ChatGPT (discussed further in Chapter 17), are one popular example of this type of model.

Similar to generating one word at a time, in the context of image generation, autoregressive models like PixelCNN try to predict one pixel at a time, given the pixels they have seen so far. Such a model might predict pixels from top left to bottom right, in a raster scan order, or in any other defined order.

To illustrate how autoregressive models generate an image one pixel at a time, suppose we have an image of size $H \times W$ (where H is the height and W is the width), ignoring the color channel for simplicity's sake. This image consists of N pixels, where $i = 1, \ldots, N$. The probability of observing a particular image in the dataset is then $P(Image) = P(i_1, i_2, \ldots, i_N)$. Based on the chain rule of probability in statistics, we can decompose this joint probability into conditional probabilities:

$$P(Image) = P\left(i_1, i_2, \ldots, i_N\right)$$
$$= P\left(i_1\right) \cdot P\left(i_2 \mid i_1\right) \cdot P\left(i_3 \mid i_1, i_2\right) \ldots P\left(i_N \mid i_1 \text{ to } i_{N-1}\right)$$

Here, $P(i_1)$ is the probability of the first pixel, $P(i_2 \mid i_1)$ is the probability of the second pixel given the first pixel, $P(i_3 \mid i_1, i_2)$ is the probability of the third pixel given the first and second pixels, and so on.

In the context of image generation, an autoregressive model essentially tries to predict one pixel at a time, as described earlier, given the pixels it has seen so far. Figure 9-5 illustrates this process, where pixels $i_1, \ldots, i_{53}$ represent the context and pixel i_{54} is the next pixel to be generated.

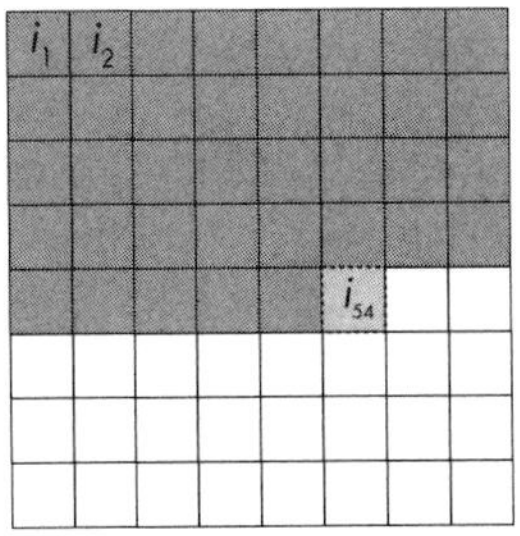

Figure 9-5: Autoregressive pixel generation

The advantage of autoregressive models is that the next-pixel (or word) prediction is relatively straightforward and interpretable. In addition, autoregressive models can compute the likelihood of data exactly, similar to flow-based models, which can be useful for tasks like anomaly detection. Furthermore, autoregressive models are easier to train than GANs as they don't suffer from issues like mode collapse and other training instabilities.

However, autoregressive models can be slow at generating new samples. This is because they have to generate data one step at a time (for example, pixel by pixel for images), which can be computationally expensive. Autoregressive models may also struggle to capture long-range dependencies because each output is conditioned only on previously generated outputs.

In terms of overall image quality, autoregressive models are therefore usually worse than GANs but are easier to train.

Diffusion Models

As discussed in the previous section, flow-based models transform a simple distribution (such as a standard normal distribution) into a complex one (the target distribution) by applying a sequence of invertible and differentiable transformations (flows). Like flow-based models, *diffusion models* also apply a series of transformations. However, the underlying concept is fundamentally different.

Diffusion models transform the input data distribution into a simple noise distribution over a series of steps using stochastic differential equations. Diffusion is a stochastic process in which noise is progressively added to the data until it resembles a simpler distribution, like Gaussian noise. To generate new samples, the process is then reversed, starting from noise and progressively removing it.

Figure 9-6 outlines the process of adding and removing Gaussian noise from an input image x. During inference, the reverse diffusion process is used to generate a new image x, starting with the noise tensor z_n sampled from a Gaussian distribution.

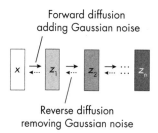

Figure 9-6: The diffusion process

While both diffusion models and flow-based models are generative models aiming to learn complex data distributions, they approach the problem from different angles. Flow-based models use deterministic invertible transformations, while diffusion models use the aforementioned stochastic diffusion process.

Recent projects have established state-of-the-art performance in generating high-quality images with realistic details and textures. Diffusion models are also easier to train than GANs. The downside of diffusion models, however, is that they are slower to sample from since they require running a series of sequential steps, similar to flow-based models and autoregressive models.

Consistency Models

Consistency models train a neural network to map a noisy image to a clean one. The network is trained on a dataset of pairs of noisy and clean images and learns to identify patterns in the clean images that are modified by noise. Once the network is trained, it can be used to generate reconstructed images from noisy images in one step.

Consistency model training employs an *ordinary differential equation (ODE)* trajectory, a path that a noisy image follows as it is gradually denoised. The ODE trajectory is defined by a set of differential equations that describe how the noise in the image changes over time, as illustrated in Figure 9-7.

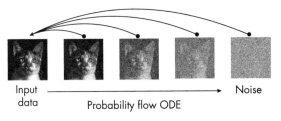

Input data ——————————————→ Noise

Probability flow ODE

Figure 9-7: Trajectories of a consistency model for image denoising

As Figure 9-7 demonstrates, we can think of consistency models as models that learn to map any point from a probability flow ODE, which smoothly converts data to noise, to the input.

At the time of writing, consistency models are the most recent type of generative AI model. Based on the original paper proposing this method, consistency models rival diffusion models in terms of image quality. Consistency models are also faster than diffusion models because they do not require an iterative process to generate images; instead, they generate images in a single step.

However, while consistency models allow for faster inference, they are still expensive to train because they require a large dataset of pairs of noisy and clean images.

Recommendations

Deep Boltzmann machines are interesting from a historical perspective since they were one of the pioneering models to effectively demonstrate the concept of unsupervised learning. Flow-based and autoregressive models may be useful when you need to estimate exact likelihoods. However, other models are usually the first choice when it comes to generating high-quality images.

In particular, VAEs and GANs have competed for years to generate the best high-fidelity images. However, in 2022, diffusion models began to take over image generation almost entirely. Consistency models are a promising alternative to diffusion models, but it remains to be seen whether they become more widely adopted to generate state-of-the-art results. The trade-off here is that sampling from diffusion models is generally slower since it involves a sequence of noise-removal steps that must be run in order, similar to autoregressive models. This can make diffusion models less practical for some applications requiring fast sampling.

Exercises

9-1. How would we evaluate the quality of the images generated by a generative AI model?

9-2. Given this chapter's description of consistency models, how would we use them to generate new images?

References

- The original paper proposing variational autoencoders: Diederik P. Kingma and Max Welling, "Auto-Encoding Variational Bayes" (2013), *https://arxiv.org/abs/1312.6114.*

- The paper introducing generative adversarial networks: Ian J. Goodfellow et al., "Generative Adversarial Networks" (2014), *https://arxiv.org/abs/1406.2661.*

- The paper introducing NICE: Laurent Dinh, David Krueger, and Yoshua Bengio, "NICE: Non-linear Independent Components Estimation" (2014), *https://arxiv.org/abs/1410.8516.*

- The paper proposing the autoregressive PixelCNN model: Aaron van den Oord et al., "Conditional Image Generation with PixelCNN Decoders" (2016), *https://arxiv.org/abs/1606.05328.*

- The paper introducing the popular Stable Diffusion latent diffusion model: Robin Rombach et al., "High-Resolution Image Synthesis with Latent Diffusion Models" (2021), *https://arxiv.org/abs/2112.10752.*

- The Stable Diffusion code implementation: *https://github.com/CompVis/stable-diffusion.*

- The paper originally proposing consistency models: Yang Song et al., "Consistency Models" (2023), *https://arxiv.org/abs/2303.01469.*

10

SOURCES OF RANDOMNESS

What are the common sources of randomness when training deep neural networks that can cause non-reproducible behavior during training and inference?

When training or using machine learning models such as deep neural networks, several sources of randomness can lead to different results every time we train or run these models, even though we use the same overall settings. Some of these effects are accidental and some are intended. The following sections categorize and discuss these various sources of randomness.

Optional hands-on examples for most of these categories are provided in the *supplementary/q10-random-sources* subfolder at *https://github.com/rasbt/ MachineLearning-QandAI-book*.

Model Weight Initialization

All common deep neural network frameworks, including TensorFlow and PyTorch, randomly initialize the weights and bias units at each layer by default. This means that the final model will be different every time we start the training. The reason these trained models will differ when we start with different random weights is the nonconvex nature of the loss, as illustrated in Figure 10-1. As the figure shows, the loss will converge to different local minima depending on where the initial starting weights are located.

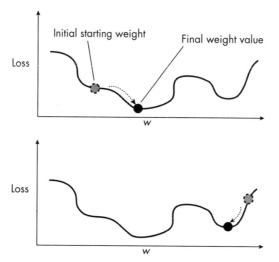

Figure 10-1: Different starting weights can lead to different final weights.

In practice, it is therefore recommended to run the training (if the computational resources permit) at least a handful of times; unlucky initial weights can sometimes cause the model not to converge or to converge to a local minimum corresponding to poorer predictive accuracy.

However, we can make the random weight initialization deterministic by seeding the random generator. For instance, if we set the seed to a specific value like 123, the weights will still initialize with small random values. Nonetheless, the neural network will consistently initialize with the same small random weights, enabling accurate reproduction of results.

Dataset Sampling and Shuffling

When we train and evaluate machine learning models, we usually start by dividing a dataset into training and test sets. This requires random sampling since we have to decide which examples we put into a training set and which examples we put into a test set.

In practice, we often use model evaluation techniques such as k-fold cross-validation or holdout validation. In holdout validation, we split the training set into training, validation, and test datasets, which are also sampling procedures influenced by randomness. Similarly, unless we use a fixed random seed, we get a different model each time we partition the dataset or tune or evaluate the model using k-fold cross-validation since the training partitions will differ.

Nondeterministic Algorithms

We may include random components and algorithms depending on the architecture and hyperparameter choices. A popular example of this is *dropout*.

Dropout works by randomly setting a fraction of a layer's units to zero during training, which helps the model learn more robust and generalized representations. This "dropping out" is typically applied at each training iteration with a probability p, a hyperparameter that controls the fraction of units dropped out. Typical values for p are in the range of 0.2 to 0.8.

To illustrate this concept, Figure 10-2 shows a small neural network where dropout randomly drops a subset of the hidden layer nodes in each forward pass during training.

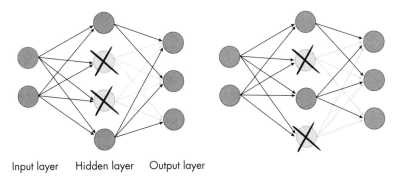

Input layer   Hidden layer   Output layer

Figure 10-2: In dropout, hidden nodes are intermittently and randomly disabled during each forward pass in training.

To create reproducible training runs, we must seed the random generator before training with dropout (analogous to seeding the random generator before initializing the model weights). During inference, we need to disable dropout to guarantee deterministic results. Each deep learning framework has a specific setting for that purpose—a PyTorch example is included in the *supplementary/q10-random-sources* subfolder at *https://github .com/rasbt/MachineLearning-QandAI-book*.

Different Runtime Algorithms

The most intuitive or simplest implementation of an algorithm or method is not always the best one to use in practice. For example, when training deep neural networks, we often use efficient alternatives and approximations to gain speed and resource advantages during training and inference.

A popular example is the convolution operation used in convolutional neural networks. There are several possible ways to implement the convolution operation:

The classic direct convolution The common implementation of discrete convolution via an element-wise product between the input and the window, followed by summing the result to get a single number. (See Chapter 12 for a discussion of the convolution operation.)

FFT-based convolution   Uses fast Fourier transform (FFT) to convert the convolution into an element-wise multiplication in the frequency domain.

Winograd-based convolution   An efficient algorithm for small filter sizes (like 3×3) that reduces the number of multiplications required for the convolution.

Different convolution algorithms have different trade-offs in terms of memory usage, computational complexity, and speed. By default, libraries such as the CUDA Deep Neural Network library (cuDNN), which are used in PyTorch and TensorFlow, can choose different algorithms for performing convolution operations when running deep neural networks on GPUs. However, the deterministic algorithm choice has to be explicitly enabled. In PyTorch, for example, this can be done by setting

```
torch.use_deterministic_algorithms(True)
```

While these approximations yield similar results, subtle numerical differences can accumulate during training and cause the training to converge to slightly different local minima.

Hardware and Drivers

Training deep neural networks on different hardware can also produce different results due to small numeric differences, even when the same algorithms are used and the same operations are executed. These differences may sometimes be due to different numeric precision for floating-point operations. However, small numeric differences may also arise due to hardware and software optimization, even at the same precision.

For instance, different hardware platforms may have specialized optimizations or libraries that can slightly alter the behavior of deep learning algorithms. To give one example of how different GPUs can produce different modeling results, the following is a quotation from the official NVIDIA documentation: "Across different architectures, no cuDNN routines guarantee bit-wise reproducibility. For example, there is no guarantee of bit-wise reproducibility when comparing the same routine run on NVIDIA Volta™ and NVIDIA Turing™ [. . .] and NVIDIA Ampere architecture."

Randomness and Generative AI

Besides the various sources of randomness mentioned earlier, certain models may also exhibit random behavior during inference that we can think of as "randomness by design." For instance, generative image and language models may create different results for identical prompts to produce a diverse sample of results. For image models, this is often so that users can select the most accurate and aesthetically pleasing image. For language models, this is often to vary the responses, for example, in chat agents, to avoid repetition.

The intended randomness in generative image models during inference is often due to sampling different noise values at each step of the reverse process. In diffusion models, a noise schedule defines the noise variance added at each step of the diffusion process.

Autoregressive LLMs like GPT tend to create different outputs for the same input prompt (GPT will be discussed at greater length in Chapters 14 and 17). The ChatGPT user interface even has a Regenerate Response button for that purpose. The ability to generate different results is due to the sampling strategies these models employ. Techniques such as top-k sampling, nucleus sampling, and temperature scaling influence the model's output by controlling the degree of randomness. This is a feature, not a bug, since it allows for diverse responses and prevents the model from producing overly deterministic or repetitive outputs. (See Chapter 9 for a more in-depth overview of generative AI and deep learning models; see Chapter 17 for more detail on autoregressive LLMs.)

*Top-*k *sampling*, illustrated in Figure 10-3, works by sampling tokens from the top k most probable candidates at each step of the next-word generation process.

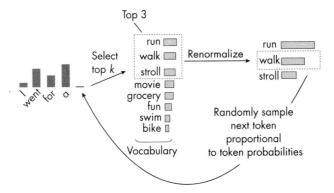

Figure 10-3: Top-k sampling

Given an input prompt, the language model produces a probability distribution over the entire vocabulary (the candidate words) for the next token. Each token in the vocabulary is assigned a probability based on the model's understanding of the context. The selected top-k tokens are then renormalized so that the probabilities sum to 1. Finally, a token is sampled from the renormalized top-k probability distribution and is appended to the input prompt. This process is repeated for the desired length of the generated text or until a stop condition is met.

Nucleus sampling (also known as *top-*p *sampling*), illustrated in Figure 10-4, is an alternative to top-k sampling.

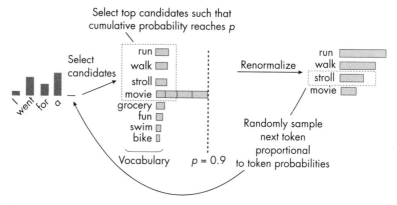

Figure 10-4: Nucleus sampling

Similar to top-k sampling, the goal of nucleus sampling is to balance diversity and coherence in the output. However, nucleus and top-k sampling differ in how to select the candidate tokens for sampling at each step of the generation process. Top-k sampling selects the k most probable tokens from the probability distribution produced by the language model, regardless of their probabilities. The value of k remains fixed throughout the generation process. Nucleus sampling, on the other hand, selects tokens based on a probability threshold p, as shown in Figure 10-4. It then accumulates the most probable tokens in descending order until their cumulative probability meets or exceeds the threshold p. In contrast to top-k sampling, the size of the candidate set (nucleus) can vary at each step.

Exercises

10-1. Suppose we train a neural network with top-k or nucleus sampling where k and p are hyperparameter choices. Can we make the model behave deterministically during inference without changing the code?

10-2. In what scenarios might random dropout behavior during inference be desired?

References

- For more about different data sampling and model evaluation techniques, see my article: "Model Evaluation, Model Selection, and Algorithm Selection in Machine Learning" (2018), *https://arxiv.org/abs/1811.12808*.

- The paper that originally proposed the dropout technique: Nitish Srivastava et al., "Dropout: A Simple Way to Prevent Neural Networks from Overfitting" (2014), *https://jmlr.org/papers/v15/srivastava14a.html*.

- A detailed paper on FFT-based convolution: Lu Chi, Borui Jiang, and Yadong Mu, "Fast Fourier Convolution" (2020), *https://dl.acm.org/doi/abs/10.5555/3495724.3496100*.

- Details on Winograd-based convolution: Syed Asad Alam et al., "Winograd Convolution for Deep Neural Networks: Efficient Point Selection" (2022), *https://arxiv.org/abs/2201.10369*.

- More information about the deterministic algorithm settings in PyTorch: *https://pytorch.org/docs/stable/generated/torch.use_deterministic_algorithms.html*.

- For details on the deterministic behavior of NVIDIA graphics cards, see the "Reproducibility" section of the official NVIDIA documentation: *https://docs.nvidia.com/deeplearning/cudnn/developer-guide/index.html#reproducibility*.

PART II

COMPUTER VISION

11

CALCULATING THE NUMBER OF PARAMETERS

How do we compute the number of parameters in a convolutional neural network, and why is this information useful?

Knowing the number of parameters in a model helps gauge the model's size, which affects storage and memory requirements. The following sections will explain how to compute the convolutional and fully connected layer parameter counts.

How to Find Parameter Counts

Suppose we are working with a convolutional network that has two convolutional layers with kernel size 5 and kernel size 3, respectively. The first convolutional layer has 3 input channels and 5 output channels, and the second one has 5 input channels and 12 output channels. The stride of these convolutional layers is 1. Furthermore, the network has two pooling layers, one with a kernel size of 3 and a stride of 2, and another with a kernel size of 5 and a stride of 2. It also has two fully connected hidden layers with 192 and 128 hidden units each, where the output layer is a classification layer for 10 classes. The architecture of this network is illustrated in Figure 11-1.

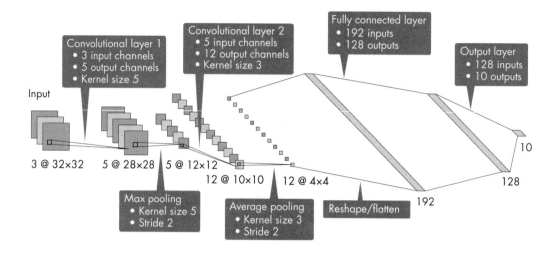

Figure 11-1: A convolutional neural network with two convolutional and two fully connected layers

What is the number of trainable parameters in this convolutional network? We can approach this problem from left to right, computing the number of parameters for each layer and then summing up these counts to obtain the total number of parameters. Each layer's number of trainable parameters consists of weights and bias units.

Convolutional Layers

In a convolutional layer, the number of weights depends on the kernel's width and height and the number of input and output channels. The number of bias units depends on the number of output channels only. To illustrate the computation step by step, suppose we have a kernel width and height of 5, one input channel, and one output channel, as illustrated in Figure 11-2.

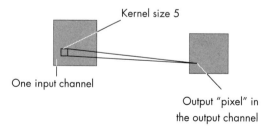

Figure 11-2: A convolutional layer with one input channel and one output channel

In this case, we have 26 parameters, since we have $5 \times 5 = 25$ weights via the kernel plus the bias unit. The computation to determine an output value or pixel z is $z = b + \sum_j w_j x_j$, where x_j represents an input pixel, w_j represents a weight parameter of the kernel, and b is the bias unit.

Now, suppose we have three input channels, as illustrated in Figure 11-3.

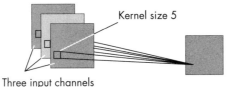

Three input channels

Figure 11-3: A convolutional layer with three
input channels and one output channel

In that case, we compute the output value by performing the aforementioned operation, $\sum_j w_j x_j$, for each input channel and then add the bias unit. For three input channels, this would involve three different kernels with three sets of weights:

$$z = \sum_j w_j^{(1)} x_j + \sum_j w_j^{(2)} x_j + \sum_j w_j^{(3)} x_j + b$$

Since we have three sets of weights ($w^{(1)}$, $w^{(2)}$, and $w^{(3)}$ for $j = [1, \dots, 25]$), we have $3 \times 25 + 1 = 76$ parameters in this convolutional layer.

We use one kernel for each output channel, where each kernel is unique to a given output channel. Thus, if we extend the number of output channels from one to five, as shown in Figure 11-4, we extend the number of parameters by a factor of 5. In other words, if the kernel for one output channel has 76 parameters, the 5 kernels required for the five output channels will have $5 \times 76 = 380$ parameters.

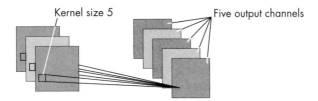

Kernel size 5   Five output channels

Figure 11-4: A convolutional layer with three input channels
and five output channels

Returning to the neural network architecture illustrated in Figure 11-1 at the beginning of this section, we compute the number of parameters in the convolutional layers based on the kernel size and number of input and output channels. For example, the first convolutional layer has three input channels, five output channels, and a kernel size of 5. Thus, its number of parameters is $5 \times (5 \times 5 \times 3) + 5 = 380$. The second convolutional layer, with five input channels, 12 output channels, and a kernel size of 3, has $12 \times (3 \times 3 \times 5) + 12 = 552$ parameters. Since the pooling layers do not have any trainable parameters, we can count $380 + 552 = 932$ for the convolutional part of this architecture.

Next, let's see how we can compute the number of parameters of fully connected layers.

Fully Connected Layers

Counting the number of parameters in a fully connected layer is relatively straightforward. A fully connected node connects each input node to each output node, so the number of weights is the number of inputs times the number of outputs plus the bias units added to the output. For example, if we have a fully connected layer with five inputs and three outputs, as shown in Figure 11-5, we have $5 \times 3 = 15$ weights and three bias units, that is, 18 parameters total.

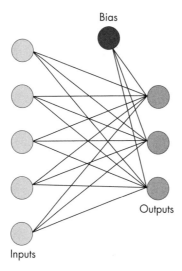

Figure 11-5: A fully connected layer with five inputs and three outputs

Returning once more to the neural network architecture illustrated in Figure 11-1, we can now calculate the parameters in the fully connected layers as follows: $192 \times 128 + 128 = 24{,}704$ in the first fully connected layer and $128 \times 10 + 10 = 1{,}290$ in the second fully connected layer, the output layer. Hence, we have $24{,}704 + 1{,}290 = 25{,}994$ in the fully connected part of this network. After adding the 932 parameters from the convolutional layers and the 25,994 parameters from the fully connected layers, we can conclude that this network's total number of parameters is 26,926.

As a bonus, interested readers can find PyTorch code to compute the number of parameters programmatically in the *supplementary/q11-conv-size* subfolder at *https://github.com/rasbt/MachineLearning-QandAI-book*.

Practical Applications

Why do we care about the number of parameters at all? First, we can use this number to estimate a model's complexity. As a rule of thumb, the more parameters there are, the more training data we'll need to train the model well.

The number of parameters also lets us estimate the size of the neural network, which in turn helps us estimate whether the network can fit into GPU memory. Although the memory requirement during training often exceeds the model size due to the additional memory required for carrying out matrix multiplications and storing gradients, model size gives us a ballpark sense of whether training the model on a given hardware setup is feasible.

Exercises

11-1. Suppose we want to optimize the neural network using a plain stochastic gradient descent (SGD) optimizer or the popular Adam optimizer. What are the respective numbers of parameters that need to be stored for SGD and Adam?

11-2. Suppose we're adding three batch normalization (BatchNorm) layers: one after the first convolutional layer, one after the second convolutional layer, and another one after the first fully connected layer (we typically do not want to add BatchNorm layers to the output layer). How many additional parameters do these three BatchNorm layers add to the model?

12

FULLY CONNECTED AND CONVOLUTIONAL LAYERS

Under which circumstances can we replace fully connected layers with convolutional layers to perform the same computation?

Replacing fully connected layers with convolutional layers can offer advantages in terms of hardware optimization, such as by utilizing specialized hardware accelerators for convolution operations. This can be particularly relevant for edge devices.

There are exactly two scenarios in which fully connected layers and convolutional layers are equivalent: when the size of the convolutional filter is equal to the size of the receptive field and when the size of the convolutional filter is 1. As an illustration of these two scenarios, consider a fully connected layer with two input and four output units, as shown in Figure 12-1.

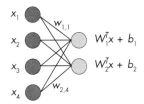

Figure 12-1: Four inputs and two outputs connected via eight weight parameters

The fully connected layer in this figure consists of eight weights and two bias units. We can compute the output nodes via the following dot products:

Node 1 $\quad w_{1,1} \times x_1 + w_{1,2} \times x_2 + w_{1,3} \times x_3 + w_{1,4} \times x_4 + b_1$

Node 2 $\quad w_{2,1} \times x_1 + w_{2,2} \times x_2 + w_{2,3} \times x_3 + w_{2,4} \times x_4 + b_2$

The following two sections illustrate scenarios in which convolutional layers can be defined to produce exactly the same computation as the fully connected layer described.

When the Kernel and Input Sizes Are Equal

Let's start with the first scenario, where the size of the convolutional filter is equal to the size of the receptive field. Recall from Chapter 11 how we compute a number of parameters in a convolutional kernel with one input channel and multiple output channels. We have a kernel size of 2×2, one input channel, and two output channels. The input size is also 2×2, a reshaped version of the four inputs depicted in Figure 12-2.

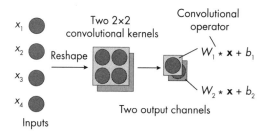

Figure 12-2: A convolutional layer with a 2×2 kernel that equals the input size and two output channels

If the convolutional kernel dimensions equal the input size, as depicted in Figure 12-2, there is no sliding window mechanism in the convolutional layer. For the first output channel, we have the following set of weights:

$$W_1 = \begin{bmatrix} w_{1,1} & w_{1,2} \\ w_{1,3} & w_{1,4} \end{bmatrix}$$

For the second output channel, we have the following set of weights:

$$W_2 = \begin{bmatrix} w_{2,1} & w_{2,2} \\ w_{2,3} & w_{2,4} \end{bmatrix}$$

If the inputs are organized as

$$x = \begin{bmatrix} x_1 & x_2 \\ x_3 & x_4 \end{bmatrix}$$

we calculate the first output channel as $o_1 = \sum_i (W_1 * \mathbf{x})_i + b_1$, where the convolutional operator $*$ is equal to an element-wise multiplication. In other words, we perform an element-wise multiplication between two matrices, W_1 and $\mathbf{x}$, and then compute the output as the sum over these elements; this equals the dot product in the fully connected layer. Lastly, we add the bias unit. The computation for the second output channel works analogously: $o_2 = \sum_i (W_2 * \mathbf{x})_i + b_2$.

As a bonus, the supplementary materials for this book include PyTorch code to show this equivalence with a hands-on example in the *supplementary/ q12-fc-cnn-equivalence* subfolder at *https://github.com/rasbt/MachineLearning -QandAI-book*.

When the Kernel Size Is 1

The second scenario assumes that we reshape the input into an input "image" with 1×1 dimensions where the number of "color channels" equals the number of input features, as depicted in Figure 12-3.

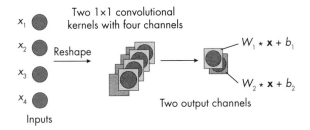

Figure 12-3: The number of output nodes equals the number of channels if the kernel size is equal to the input size.

Each kernel consists of a stack of weights equal to the number of input channels. For instance, for the first output layer, the weights are

$$W_1 = [w_1^{(1)} w_1^{(2)} w_1^{(3)} w_1^{(4)}]$$

while the weights for the second channel are:

$$W_2 = [w_2^{(1)} w_2^{(2)} w_2^{(3)} w_2^{(4)}]$$

To get a better intuitive understanding of this computation, check out the illustrations in Chapter 11, which describe how to compute the parameters in a convolutional layer.

Recommendations

The fact that fully connected layers can be implemented as equivalent convolutional layers does not have immediate performance or other advantages on standard computers. However, replacing fully connected layers with convolutional layers can offer advantages in combination with developing specialized hardware accelerators for convolution operations.

Moreover, understanding the scenarios where fully connected layers are equivalent to convolutional layers aids in understanding the mechanics of these layers. It also lets us implement convolutional neural networks without any use of fully connected layers, if desired, to simplify code implementations.

Exercises

12-1. How would increasing the stride affect the equivalence discussed in this chapter?

12-2. Does padding affect the equivalence between fully connected layers and convolutional layers?

13

LARGE TRAINING SETS FOR VISION TRANSFORMERS

Why do vision transformers (ViTs) generally require larger training sets than convolutional neural networks (CNNs)?

Each machine learning algorithm and model encodes a particular set of assumptions or prior knowledge, commonly referred to as *inductive biases*, in its design. Some inductive biases are workarounds to make algorithms computationally more feasible, other inductive biases are based on domain knowledge, and some inductive biases are both.

CNNs and ViTs can be used for the same tasks, including image classification, object detection, and image segmentation. CNNs are mainly composed of convolutional layers, while ViTs consist primarily of multi-head attention blocks (discussed in Chapter 8 in the context of transformers for natural language inputs).

CNNs have more inductive biases that are hardcoded as part of the algorithmic design, so they generally require less training data than ViTs. In a sense, ViTs are given more degrees of freedom and can or must learn certain inductive biases from the data (assuming that these biases are conducive to optimizing the training objective). However, everything that needs to be learned requires more training examples.

The following sections explain the main inductive biases encountered in CNNs and how ViTs work well without them.

Inductive Biases in CNNs

The following are the primary inductive biases that largely define how CNNs function:

Local connectivity   In CNNs, each unit in a hidden layer is connected to only a subset of neurons in the previous layer. We can justify this restriction by assuming that neighboring pixels are more relevant to each other than pixels that are farther apart. As an intuitive example, consider how this assumption applies to the context of recognizing edges or contours in an image.

Weight sharing   Via the convolutional layers, we use the same small set of weights (the kernels or filters) throughout the whole image. This reflects the assumption that the same filters are useful for detecting the same patterns in different parts of the image.

Hierarchical processing   CNNs consist of multiple convolutional layers to extract features from the input image. As the network progresses from the input to the output layers, low-level features are successively combined to form increasingly complex features, ultimately leading to the recognition of more complex objects and shapes. Furthermore, the convolutional filters in these layers learn to detect specific patterns and features at different levels of abstraction.

Spatial invariance   CNNs exhibit the mathematical property of spatial invariance, meaning the output of a model remains consistent even if the input signal is shifted to a different location within the spatial domain. This characteristic arises from the combination of local connectivity, weight sharing, and the hierarchical architecture mentioned earlier.

The combination of local connectivity, weight sharing, and hierarchical processing in a CNN leads to spatial invariance, allowing the model to recognize the same pattern or feature regardless of its location in the input image.

Translation invariance is a specific case of spatial invariance in which the output remains the same after a shift or translation of the input signal in the spatial domain. In this context, the emphasis is solely on moving an object to a different location within an image without any rotations or alterations of its other attributes.

In reality, convolutional layers and networks are not truly translation-invariant; rather, they achieve a certain level of translation equivariance. What is the difference between translation invariance and equivariance? *Translation invariance* means that the output does not change with an input shift, while *translation equivariance* implies that the output shifts with the input in a corresponding manner. In other words, if we shift the input object to the right, the results will correspondingly shift to the right, as illustrated in Figure 13-1.

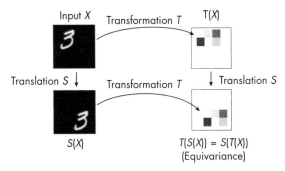

Figure 13-1: Equivariance under different image translations

As Figure 13-1 shows, under translation invariance, we get the same output pattern regardless of the order in which we apply the operations: transformation followed by translation or translation followed by transformation.

As mentioned earlier, CNNs achieve translation equivariance through a combination of their local connectivity, weight sharing, and hierarchical processing properties. Figure 13-2 depicts a convolutional operation to illustrate the local connectivity and weight-sharing priors. This figure demonstrates the concept of translation equivariance in CNNs, in which a convolutional filter captures the input signal (the two dark blocks) irrespective of where it is located in the input.

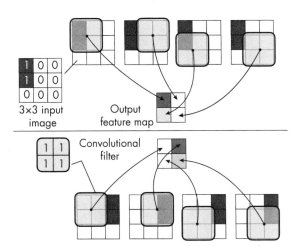

Figure 13-2: Convolutional filters and translation equivariance

Figure 13-2 shows a 3×3 input image that consists of two nonzero pixel values in the upper-left corner (top portion of the figure) or upper-right corner (bottom portion of the figure). If we apply a 2×2 convolutional filter to these two input image scenarios, we can see that the output feature maps contain the same extracted pattern, which is on either the left (top of the figure) or the right (bottom of the figure), demonstrating the translation equivariance of the convolutional operation.

For comparison, a fully connected network such as a multilayer percep-tron lacks this spatial invariance or equivariance. To illustrate this point, picture a multilayer perceptron with one hidden layer. Each pixel in the in-put image is connected with each value in the resulting output. If we shift the input by one or more pixels, a different set of weights will be activated, as illustrated in Figure 13-3.

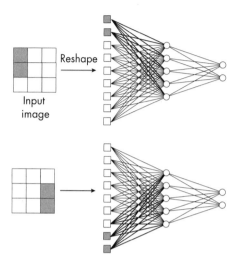

Figure 13-3: Location-specific weights in fully connected layers

Like fully connected networks, ViT architecture (and transformer ar-chitecture in general) lacks the inductive bias for spatial invariance or equi-variance. For instance, the model produces different outputs if we place the same object in two different spatial locations within an image. This is not ideal, as the semantic meaning of an object (the concept that an object rep-resents or conveys) remains the same based on its location. Consequently, it must learn these invariances directly from the data. To facilitate learning useful patterns present in CNNs requires pretraining over a larger dataset.

A common workaround for adding positional information in ViTs is to use relative positional embeddings (also known as *relative positional encod-ings*) that consider the relative distance between two tokens in the input se-quence. However, while relative embeddings encode information that helps transformers keep track of the relative location of tokens, the transformer still needs to learn from the data whether and how far spatial information is relevant for the task at hand.

ViTs Can Outperform CNNs

The hardcoded assumptions via the inductive biases discussed in previous sections reduce the number of parameters in CNNs substantially compared to fully connected layers. On the other hand, ViTs tend to have larger num-bers of parameters than CNNs, which require more training data. (Refer to Chapter 11 for a refresher on how to precisely calculate the number of parameters in fully connected and convolutional layers.)

ViTs may underperform compared to popular CNN architectures without extensive pretraining, but they can perform very well with a sufficiently large pretraining dataset. In contrast to language transformers, where unsupervised pretraining (such as self-supervised learning, discussed in Chapter 2) is a preferred choice, vision transformers are often pretrained using large, labeled datasets like ImageNet, which provides millions of labeled images for training, and regular supervised learning.

An example of ViTs surpassing the predictive performance of CNNs, given enough data, can be observed from initial research on the ViT architecture, as shown in the paper "An Image Is Worth 16x16 Words: Transformers for Image Recognition at Scale." This study compared ResNet, a type of convolutional network, with the original ViT design using different dataset sizes for pretraining. The findings also showed that the ViT model excelled over the convolutional approach only after being pretrained on a minimum of 100 million images.

Inductive Biases in ViTs

ViTs also possess some inductive biases. For example, vision transformers *patchify* the input image to process each input patch individually. Here, each patch can attend to all other patches so that the model learns relationships between far-apart patches in the input image, as illustrated in Figure 13-4.

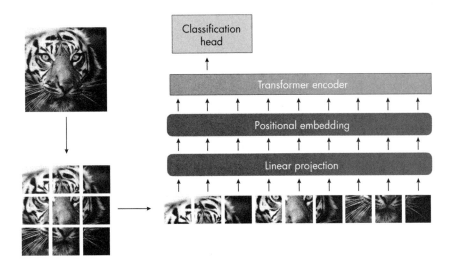

Figure 13-4: How a vision transformer operates on image patches

The patchify inductive bias allows ViTs to scale to larger image sizes without increasing the number of parameters in the model, which can be computationally expensive. By processing smaller patches individually, ViTs can efficiently capture spatial relationships between image regions while benefiting from the global context captured by the self-attention mechanism.

This raises another question: how and what do ViTs learn from the training data? ViTs learn more uniform feature representations across all layers, with self-attention mechanisms enabling early aggregation of global information. In addition, the residual connections in ViTs strongly propagate features from lower to higher layers, in contrast to the more hierarchical structure of CNNs.

ViTs tend to focus more on global than local relationships because their self-attention mechanism allows the model to consider long-range dependencies between different parts of the input image. Consequently, the self-attention layers in ViTs are often considered low-pass filters that focus more on shapes and curvature.

In contrast, the convolutional layers in CNNs are often considered high-pass filters that focus more on texture. However, keep in mind that convolutional layers can act as both high-pass and low-pass filters, depending on the learned filters at each layer. High-pass filters detect an image's edges, fine details, and texture, while low-pass filters capture more global, smooth features and shapes. CNNs achieve this by applying convolutional kernels of varying sizes and learning different filters at each layer.

Recommendations

ViTs have recently begun outperforming CNNs if enough data is available for pretraining. However, this doesn't make CNNs obsolete, as methods such as the popular EfficientNetV2 CNN architecture are less memory and data hungry.

Moreover, recent ViT architectures don't rely solely on large datasets, parameter numbers, and self-attention. Instead, they have taken inspiration from CNNs and added soft convolutional inductive biases or even complete convolutional layers to get the best of both worlds.

In short, vision transformer architectures without convolutional layers generally have fewer spatial and locality inductive biases than convolutional neural networks. Consequently, vision transformers need to learn data-related concepts such as local relationships among pixels. Thus, vision transformers require more training data to achieve good predictive performance and produce acceptable visual representations in generative modeling contexts.

Exercises

13-1. Consider the patchification of the input images shown in Figure 13-4. The size of the resulting patches controls a computational and predictive performance trade-off. The optimal patch size depends on the application and desired trade-off between computational cost and model performance. Do smaller patches typically result in higher or lower computational costs?

13-2. Following up on the previous question, do smaller patches typically lead to a higher or lower prediction accuracy?

References

- The paper proposing the original vision transformer model: Alexey Dosovitskiy et al., "An Image Is Worth 16x16 Words: Transformers for Image Recognition at Scale" (2020), *https://arxiv.org/abs/2010 .11929*.

- A workaround for adding positional information in ViTs is to use relative positional embeddings: Peter Shaw, Jakob Uszkoreit, and Ashish Vaswani, "Self-Attention with Relative Position Representations" (2018), *https://arxiv.org/abs/1803.02155*.

- Residual connections in ViTs strongly propagate features from lower to higher layers, in contrast to the more hierarchical structure of CNNs: Maithra Raghu et al., "Do Vision Transformers See Like Convolutional Neural Networks?" (2021), *https://arxiv.org/abs/2108 .08810*.

- A detailed research article covering the EfficientNetV2 CNN architecture: Mingxing Tan and Quoc V. Le, "EfficientNetV2: Smaller Models and Faster Training" (2021), *https://arxiv.org/abs/2104.00298*.

- A ViT architecture that also incorporates convolutional layers: Stéphane d'Ascoli et al., "ConViT: Improving Vision Transformers with Soft Convolutional Inductive Biases" (2021), *https://arxiv .org/abs/2103.10697*.

- Another example of a ViT using convolutional layers: Haiping Wu et al., "CvT: Introducing Convolutions to Vision Transformers" (2021), *https://arxiv.org/abs/2103.15808*.

PART III

NATURAL LANGUAGE PROCESSING

14

THE DISTRIBUTIONAL HYPOTHESIS

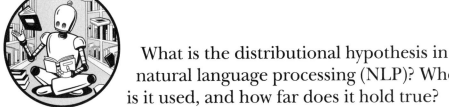

What is the distributional hypothesis in natural language processing (NLP)? Where is it used, and how far does it hold true?

The distributional hypothesis is a linguistic theory suggesting that words occurring in the same contexts tend to have similar meanings, according to the original source, "Distributional Structure" by Zellig S. Harris. Succinctly, the more similar the meanings of two words are, the more often they appear in similar contexts.

Consider the sentence in Figure 14-1, for example. The words *cats* and *dogs* often occur in similar contexts, and we could replace *cats* with *dogs* without making the sentence sound awkward. We could also replace *cats* with *hamsters*, since both are mammals and pets, and the sentence would still sound plausible. However, replacing *cats* with an unrelated word such as *sandwiches* would render the sentence clearly wrong, and replacing *cats* with the unrelated word *driving* would also make the sentence grammatically incorrect.

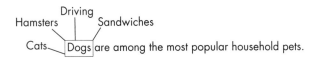

Figure 14-1: Common and uncommon words in a given context

It is easy to construct counterexamples using polysemous words, that is, words that have multiple meanings that are related but not identical. For example, consider the word *bank*. As a noun, it can refer to a financial institution, the "rising ground bordering a river," the "steep incline of a hill," or a "protective cushioning rim" (according to the Merriam-Webster dictionary). It can even be a verb: to bank on something means to rely or depend on it. These different meanings have different distributional properties and may not always occur in similar contexts.

Nonetheless, the distributional hypothesis is quite useful. Word embeddings (introduced in Chapter 1) such as Word2vec, as well as many large language transformer models, rely on this idea. This includes the masked language model in BERT and the next-word pretraining task used in GPT.

Word2vec, BERT, and GPT

The Word2vec approach uses a simple, two-layer neural network to encode words into embedding vectors such that the embedding vectors of similar words are both semantically and syntactically close. There are two ways to train a Word2vec model: the continuous bag-of-words (CBOW) approach and the skip-gram approach. When using CBOW, the Word2vec model learns to predict the current words by using the surrounding context words. Conversely, in the skip-gram model, Word2vec predicts the context words from a selected word. While skip-gram is more effective for infrequent words, CBOW is usually faster to train.

After training, word embeddings are placed within the vector space so that words with common contexts in the corpus—that is, words with semantic and syntactic similarities—are positioned close to each other, as illustrated in Figure 14-2. Conversely, dissimilar words are located farther apart in the embedding space.

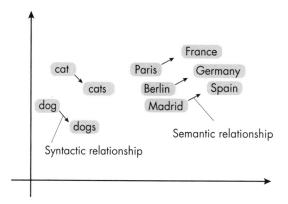

Figure 14-2: Word2vec embeddings in a two-dimensional vector space

BERT is an LLM based on the transformer architecture (see Chapter 8) that uses a masked language modeling approach that involves masking (hiding) some of the words in a sentence. Its task is to predict these masked words based on the other words in the sequence, as illustrated in Figure 14-3. This is a form of the self-supervised learning used to pretrain LLMs (see Chapter 2 for more on self-supervised learning). The pretrained model produces embeddings in which similar words (or tokens) are close in the embedding space.

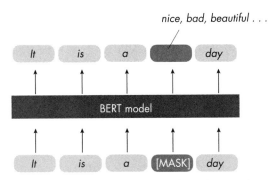

Figure 14-3: BERT's pretraining task involves predicting randomly masked words.

GPT, which like BERT is also an LLM based on the transformer architecture, functions as a decoder. Decoder-style models like GPT learn to predict subsequent words in a sequence based on the preceding ones, as illustrated in Figure 14-4. GPT contrasts with BERT, an encoder model, as it emphasizes predicting what follows rather than encoding the entire sequence simultaneously.

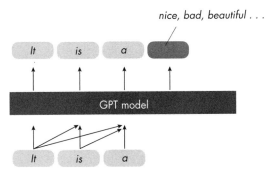

Figure 14-4: GPT is pretrained by predicting the next word.

Where BERT is a bidirectional language model that considers the whole input sequence, GPT only strictly parses previous sequence elements. This means BERT is usually better suited for classification tasks, whereas GPT is more suited for text generation tasks. Similar to BERT, GPT produces high-quality contextualized word embeddings that capture semantic similarity.

Does the Hypothesis Hold?

For large datasets, the distributional hypothesis more or less holds true, making it quite useful for understanding and modeling language patterns, word relationships, and semantic meanings. For example, this concept enables techniques like word embedding and semantic analysis, which, in turn, facilitate natural language processing tasks such as text classification, sentiment analysis, and machine translation.

In conclusion, while there are counterexamples in which the distributional hypothesis does not hold, it is a very useful concept that forms the cornerstone of modern language transformer models.

Exercises

14-1. Does the distributional hypothesis hold true in the case of homophones, or words that sound the same but have different meanings, such as *there* and *their*?

14-2. Can you think of another domain where a concept similar to the distributional hypothesis applies? (Hint: think of other input modalities for neural networks.)

References

- The original source describing the distributional hypothesis: Zellig S. Harris, "Distributional Structure" (1954), *https://doi.org/10.1080/00437956.1954.11659520*.

- The paper introducing the Word2vec model: Tomas Mikolov et al., "Efficient Estimation of Word Representations in Vector Space" (2013), *https://arxiv.org/abs/1301.3781*.

- The paper introducing the BERT model: Jacob Devlin et al., "BERT: Pre-training of Deep Bidirectional Transformers for Language Understanding" (2018), *https://arxiv.org/abs/1810.04805*.

- The paper introducing the GPT model: Alec Radford and Karthik Narasimhan, "Improving Language Understanding by Generative Pre-Training" (2018), *https://www.semanticscholar.org/paper/Improving-Language-Understanding-by-Generative-Radford-Narasimhan/cd18800a0fe0b668a1cc19f2ec95b5003d0a5035*.

- BERT produces embeddings in which similar words (or tokens) are close in the embedding space: Nelson F. Liu et al., "Linguistic Knowledge and Transferability of Contextual Representations" (2019), *https://arxiv.org/abs/1903.08855*.

- The paper showing that GPT produces high-quality contextualized word embeddings that capture semantic similarity: Fabio Petroni et al., "Language Models as Knowledge Bases?" (2019), *https://arxiv.org/abs/1909.01066*.

15

DATA AUGMENTATION FOR TEXT

How is data augmentation useful, and what are the most common augmentation techniques for text data?

Data augmentation is useful for artificially increasing dataset sizes to improve model performance, such as by reducing the degree of overfitting, as discussed in Chapter 5. This includes techniques often used in computer vision models, like rotation, scaling, and flipping.

Similarly, there are several techniques for augmenting text data. The most common include synonym replacement, word deletion, word position swapping, sentence shuffling, noise injection, back translation, and text generated by LLMs. This chapter discusses each of these, with optional code examples in the *supplementary/q15-text-augment* subfolder at *https://github.com/rasbt/MachineLearning-QandAI-book*.

Synonym Replacement

In *synonym replacement*, we randomly choose words in a sentence—often nouns, verbs, adjectives, and adverbs—and replace them with synonyms. For example, we might begin with the sentence "The cat quickly jumped over the lazy dog," and then augment the sentence as follows: "The cat rapidly jumped over the idle dog."

Synonym replacement can help the model learn that different words can have similar meanings, thereby improving its ability to understand and generate text. In practice, synonym replacement often relies on a thesaurus such as WordNet. However, using this technique requires care, as not all synonyms are interchangeable in all contexts. Most automatic text replacement tools have settings for adjusting replacement frequency and similarity thresholds. However, automatic synonym replacement is not perfect, and you might want to apply post-processing checks to filter out replacements that might not make sense.

Word Deletion

Word deletion is another data augmentation technique to help models learn. Unlike synonym replacement, which alters the text by substituting words with their synonyms, word deletion involves removing certain words from the text to create new variants while trying to maintain the overall meaning of the sentence. For example, we might begin with the sentence "The cat quickly jumped over the lazy dog" and then remove the word *quickly*: "The cat jumped over the lazy dog."

By randomly deleting words in the training data, we teach the model to make accurate predictions even when some information is missing. This can make the model more robust when encountering incomplete or noisy data in real-world scenarios. Also, by deleting nonessential words, we may teach the model to focus on key aspects of the text that are most relevant to the task at hand.

However, we must be careful not to remove critical words that may significantly alter a sentence's meaning. For example, it would be suboptimal to remove the word *cat* in the previous sentence: "The quickly jumped over the lazy dog." We must also choose the deletion rate carefully to ensure that the text still makes sense after words have been removed. Typical deletion rates might range from 10 percent to 20 percent, but this is a general guideline and could vary significantly based on the specific use case.

Word Position Swapping

In *word position swapping*, also known as *word shuffling* or *permutation*, the positions of words in a sentence are swapped or rearranged to create new versions of the sentence. If we begin with "The cat quickly jumped over the lazy dog," we might swap the positions of some words to get the following: "Quickly the cat jumped the over lazy dog."

While these sentences may sound grammatically incorrect or strange in English, they provide valuable training information for data augmentation because the model can still recognize the important words and their associations with each other. However, this method has its limitations. For example, shuffling words too much or in certain ways can drastically change the meaning of a sentence or make it completely nonsensical. Moreover, word shuffling may interfere with the model's learning process, as the positional relationships between certain words can be vital in these contexts.

Sentence Shuffling

In *sentence shuffling*, entire sentences within a paragraph or a document are rearranged to create new versions of the input text. By shuffling sentences within a document, we expose the model to different arrangements of the same content, helping it learn to recognize thematic elements and key concepts rather than relying on specific sentence order. This promotes a more robust understanding of the document's overall topic or category. Consequently, this technique is particularly useful for tasks that deal with document-level analysis or paragraph-level understanding, such as document classification, topic modeling, or text summarization.

In contrast to the aforementioned word-based methods (word position swapping, word deletion, and synonym replacement), sentence shuffling maintains the internal structure of individual sentences. This avoids the problem of altering word choice or order such that sentences become grammatically incorrect or change meaning entirely.

Sentence shuffling is useful when the order of sentences is not crucial to the overall meaning of the text. Still, it may not work well if the sentences are logically or chronologically connected. For example, consider the following paragraph: "I went to the supermarket. Then I bought ingredients to make pizza. Afterward, I made some delicious pizza." Reshuffling these sentences as follows disrupts the logical and temporal progression of the narrative: "Afterward, I made some delicious pizza. Then I bought ingredients to make pizza. I went to the supermarket."

Noise Injection

Noise injection is an umbrella term for techniques used to alter text in various ways and create variation in the texts. It may refer either to the methods described in the previous sections or to character-level techniques such as inserting random letters, characters, or typos, as shown in the following examples:

Random character insertion   "The cat qzuickly jumped over the lazy dog." (Inserted a *z* in the word *quickly*.)

Random character deletion   "The cat quickl jumped over the lazy dog." (Deleted *y* from the word *quickly*.)

Typo introduction   "The cat qickuly jumped over the lazy dog." (Introduced a typo in *quickly*, changing it to *qickuly*.)

These modifications are beneficial for tasks that involve spell-checking and text correction, but they can also help make the model more robust to imperfect inputs.

Back Translation

Back translation is one of the most widely used techniques to create variation in texts. Here, a sentence is first translated from the original language into one or more different languages, and then it is translated back into the original language. Translating back and forth often results in sentences that are semantically similar to the original sentence but have slight variations in structure, vocabulary, or grammar. This generates additional, diverse examples for training without altering the overall meaning.

For example, say we translate "The cat quickly jumped over the lazy dog" into German. We might get "Die Katze sprang schnell über den faulen Hund." We could then translate this German sentence back into English to get "The cat jumped quickly over the lazy dog."

The degree to which a sentence changes through back translation depends on the languages used and the specifics of the machine translation model. In this example, the sentence remains very similar. However, in other cases or with other languages, you might see more significant changes in wording or sentence structure while maintaining the same overall meaning.

This method requires access to reliable machine translation models or services, and care must be taken to ensure that the back-translated sentences retain the essential meaning of the original sentences.

Synthetic Data

Synthetic data generation is an umbrella term that describes methods and techniques used to create artificial data that mimics or replicates the structure of real-world data. All methods discussed in this chapter can be considered synthetic data generation techniques since they generate new data by making small changes to existing data, thus maintaining the overall meaning while creating something new.

Modern techniques to generate synthetic data now also include using decoder-style LLMs such as GPT (decoder-style LLMs are discussed in more detail in Chapter 17). We can use these models to generate new data from scratch by using "complete the sentence" or "generate example sentences" prompts, among others. We can also use LLMs as alternatives to back translation, prompting them to rewrite sentences as shown in Figure 15-1.

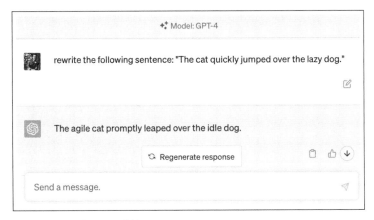

Figure 15-1: Using an LLM to rewrite a sentence

Note that an LLM, as shown in Figure 15-1, runs in a nondeterministic mode by default, which means we can prompt it multiple times to obtain a variety of rewritten sentences.

Recommendations

The data augmentation techniques discussed in this chapter are commonly used in text classification, sentiment analysis, and other NLP tasks where the amount of available labeled data might be limited.

LLMs are usually pretrained on such a vast and diverse dataset that they may not rely on these augmentation techniques as extensively as in other, more specific NLP tasks. This is because LLMs aim to capture the statistical properties of the language, and the vast amount of data on which they are trained often provides a sufficient variety of contexts and expressions. However, in the fine-tuning stages of LLMs, where a pretrained model is adapted to a specific task with a smaller, task-specific dataset, data augmentation techniques might become more relevant again, mainly if the task-specific labeled dataset size is limited.

Exercises

15-1. Can the use of text data augmentation help with privacy concerns?

15-2. What are some instances where data augmentation may not be beneficial for a specific task?

References

- The WordNet thesaurus: George A. Miller, "WordNet: A Lexical Database for English" (1995), *https://dl.acm.org/doi/10.1145/219717 .219748.*

16

SELF-ATTENTION

Where does self-attention get its name, and how is it different from previously developed attention mechanisms?

Self-attention enables a neural network to refer to other portions of the input while focusing on a particular segment, essentially allowing each part the ability to "attend" to the whole input. The original attention mechanism developed for recurrent neural networks (RNNs) is applied between two different sequences: the encoder and the decoder embeddings. Since the attention mechanisms used in transformer-based large language models is designed to work on all elements of the same set, it is known as *self*-attention.

This chapter first discusses an earlier attention mechanism developed for RNNs, the Bahdanau mechanism, in order to illustrate the motivation behind developing attention mechanism. We then compare the Bahdanau mechanism to the self-attention mechanism prevalent in transformer architectures today.

Attention in RNNs

One example of an attention mechanism used in RNNs to handle long sequences is *Bahdanau attention*. Bahdanau attention was developed to make machine learning models, particularly those used in translating languages, better at understanding long sentences. Before this type of attention, the

whole input (such as a sentence in English) was squashed into a single chunk of information, and important details could get lost,
especially if the sentence was long.

To understand the difference between regular attention and self-attention, let's begin with the illustration of the Bahdanau attention mechanism in Figure 16-1.

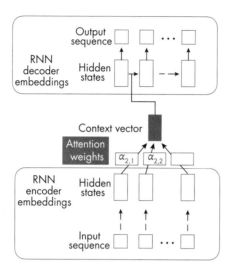

Figure 16-1: The Bahdanau mechanism uses a separate RNN to compute attention weights.

In Figure 16-1, the α values represent the attention weights for the second sequence element and each other element in the sequence from 1 to T. Furthermore, this original attention mechanism involves two RNNs. The RNN at the bottom, computing the attention weights, represents the encoder, while the RNN at the top, producing the output sequence, is a decoder.

In short, the original attention mechanism developed for RNNs is applied between two different sequences: the encoder and decoder embeddings. For each generated output sequence element, the decoder RNN at the top is based on a hidden state plus a context vector generated by the encoder. This context vector involves *all* elements of the input sequence and is a weighted sum of all input elements where the attention scores (α's) represent the weighting coefficients. This allows the decoder to access all input sequence elements (the context) at each step. The key idea is that the attention weights (and context) may differ and change dynamically at each step.

The motivation behind this complicated encoder-decoder design is that we cannot translate sentences word by word. This would result in grammatically incorrect outputs, as illustrated by the RNN architecture (a) in Figure 16-2.

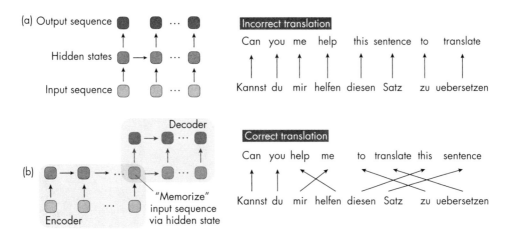

Figure 16-2: Two RNN architecture designs for translating text

Figure 16-2 shows two different sequence-to-sequence RNN designs for sentence translation. Figure 16-2(a) represents a regular sequence-to-sequence RNN that may be used to translate a sentence from German to English word by word. Figure 16-2(b) depicts an encoder-decoder RNN that first reads the whole sentence before translating it.

RNN architecture (a) is best suited for time series tasks in which we want to make one prediction at a time, such as predicting a given stock price day by day. For tasks like language translation, we typically opt for an encoder-decoder RNN, as in architecture (b) in Figure 16-2. Here, the RNN encodes the input sentence, stores it in an intermediate hidden representation, and generates the output sentence. However, this creates a bottleneck where the RNN has to memorize the whole input sentence via a single hidden state, which does not work well for longer sequences.

The bottleneck depicted in architecture (b) prompted the Bahdanau attention mechanism's original design, allowing the decoder to access all elements in the input sentence at each time step. The attention scores also give different weights to the different input elements depending on the current word that the decoder generates. For example, when generating the word *help* in the output sequence, the word *helfen* in the German input sentence may get a large attention weight, as it's highly relevant in this context.

The Self-Attention Mechanism

The Bahdanau attention mechanism relies on a somewhat complicated encoder-decoder design to model long-term dependencies in sequence-to-sequence language modeling tasks. Approximately three years after the Bahdanau mechanism, researchers worked on simplifying sequence-to-sequence modeling architectures by asking whether the RNN backbone was even needed to achieve good language translation performance. This led to the design of the original transformer architecture and self-attention mechanism.

In self-attention, the attention mechanism is applied between all elements in the same sequence (as opposed to involving two sequences), as depicted in the simplified attention mechanism in Figure 16-3. Similar to the attention mechanism for RNNs, the context vector is an attention-weighted sum over the input sequence elements.

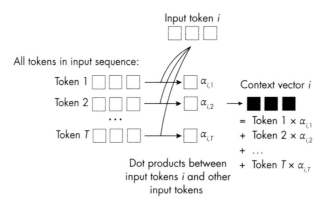

Figure 16-3: A simple self-attention mechanism without weight matrices

While Figure 16-3 doesn't include weight matrices, the self-attention mechanism used in transformers typically involves multiple weight matrices to compute the attention weights.

This chapter laid the groundwork for understanding the inner workings of transformer models and the attention mechanism. The next chapter covers the different types of transformer architectures in more detail.

Exercises

16-1. Considering that self-attention compares each sequence element with itself, what is the time and memory complexity of self-attention?

16-2. We discussed self-attention in the context of natural language processing. Could this mechanism be useful for computer vision applications as well?

References

- The paper introducing the original self-attention mechanism, also known as *scaled dot-product* attention: Ashish Vaswani et al., "Attention Is All You Need" (2017), *https://arxiv.org/abs/1706.03762*.

- The Bahdanau attention mechanism for RNNs: Dzmitry Bahdanau, Kyunghyun Cho, and Yoshua Bengio, "Neural Machine Translation by Jointly Learning to Align and Translate" (2014), *https://arxiv.org/abs/1409.0473*.

- For more about the parameterized self-attention mechanism, check out my blog post: "Understanding and Coding the Self-Attention Mechanism of Large Language Models from Scratch" at *https://sebastianraschka.com/blog/2023/self-attention-from-scratch.html*.

17

ENCODER- AND DECODER-STYLE TRANSFORMERS

What are the differences between encoder- and decoder-based language transformers? Both encoder- and decoder-style architectures use the same self-attention layers to encode word tokens. The main difference is that encoders are designed to learn embeddings that can be used for various predictive modeling tasks such as classification. In contrast, decoders are designed to generate new texts, for example, to answer user queries.

This chapter starts by describing the original transformer architecture consisting of an encoder that processes input text and a decoder that produces translations. The subsequent sections then describe how models like BERT and RoBERTa utilize only the encoder to understand context and how the GPT architectures emphasize decoder-only mechanisms for text generation.

The Original Transformer

The original transformer architecture introduced in Chapter 16 was developed for English-to-French and English-to-German language translation. It utilized both an encoder and a decoder, as illustrated in Figure 17-1.

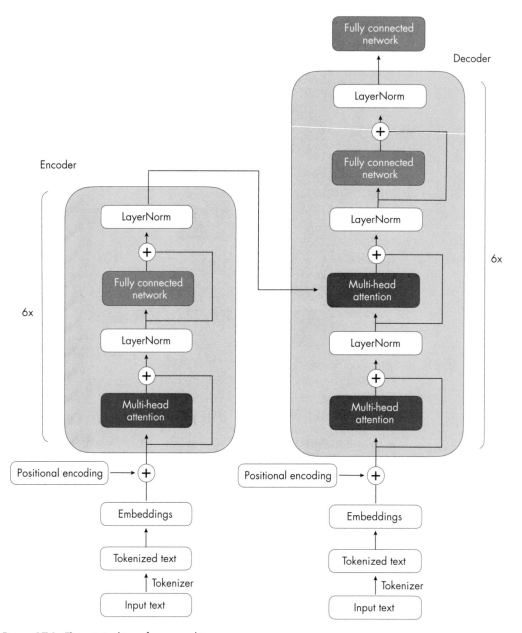

Figure 17-1: The original transformer architecture

In Figure 17-1, the input text (that is, the sentences of the text to be translated) is first tokenized into individual word tokens, which are then encoded via an embedding layer before they enter the encoder part (see Chapter 1 for more on embeddings). After a positional encoding vector is added to each embedded word, the embeddings go through a multi-head self-attention layer. This layer is followed by an addition step, indicated by a plus sign (+) in Figure 17-1, which performs a layer normalization and adds the original embeddings via a skip connection, also known as a *residual* or *shortcut* connection. Following this is a LayerNorm block, short for *layer normalization*, which normalizes the activations of the previous layer to improve the stability of the neural network's training. The addition of the original embeddings and the layer normalization steps are often summarized as the *Add & Norm step*. Finally, after entering the fully connected network—a small, multilayer perceptron consisting of two fully connected layers with a nonlinear activation function in between—the outputs are again added and normalized before they are passed to a multi-head self-attention layer of the decoder.

The decoder in Figure 17-1 has a similar overall structure to the encoder. The key difference is that the inputs and outputs are different: the encoder receives the input text to be translated, while the decoder generates the translated text.

Encoders

The encoder part in the original transformer, as illustrated in Figure 17-1, is responsible for understanding and extracting the relevant information from the input text. It then outputs a continuous representation (embedding) of the input text, which is passed to the decoder. Finally, the decoder generates the translated text (target language) based on the continuous representation received from the encoder.

Over the years, various encoder-only architectures have been developed based on the encoder module of the original transformer model outlined earlier. One notable example is BERT, which stands for bidirectional encoder representations from transformers.

As noted in Chapter 14, BERT is an encoder-only architecture based on the transformer's encoder module. The BERT model is pretrained on a large text corpus using masked language modeling and next-sentence prediction tasks. Figure 17-2 illustrates the masked language modeling pretraining objective used in BERT-style transformers.

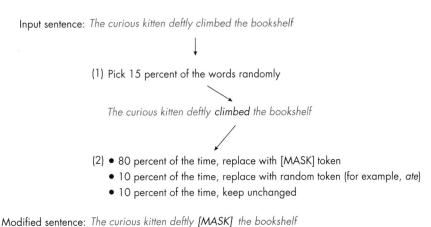

Input sentence: *The curious kitten deftly climbed the bookshelf*

(1) Pick 15 percent of the words randomly

*The curious kitten deftly **climbed** the bookshelf*

(2) • 80 percent of the time, replace with [MASK] token
    • 10 percent of the time, replace with random token (for example, *ate*)
    • 10 percent of the time, keep unchanged

Modified sentence: *The curious kitten deftly [MASK] the bookshelf*

Figure 17-2: BERT randomly masks 15 percent of the input tokens during pretraining.

As Figure 17-2 demonstrates, the main idea behind masked language modeling is to mask (or replace) random word tokens in the input sequence and then train the model to predict the original masked tokens based on the surrounding context.

In addition to the masked language modeling pretraining task illustrated in Figure 17-2, the next-sentence prediction task asks the model to predict whether the original document's sentence order of two randomly shuffled sentences is correct. For example, say that two sentences, in random order, are separated by the [SEP] token (*SEP* is short for *separate*). The brackets are a part of the token's notation and are used to make it clear that this is a special token as opposed to a regular word in the text. BERT-style transformers also use a [CLS] token. The [CLS] token serves as a placeholder token for the model, prompting the model to return a *True* or *False* label indicating whether the sentences are in the correct order:

- "[CLS] Toast is a simple yet delicious food. [SEP] It's often served with butter, jam, or honey."

- "[CLS] It's often served with butter, jam, or honey. [SEP] Toast is a simple yet delicious food."

The masked language and next-sentence pretraining objectives allow BERT to learn rich contextual representations of the input texts, which can then be fine-tuned for various downstream tasks like sentiment analysis, question answering, and named entity recognition. It's worth noting that this pretraining is a form of self-supervised learning (see Chapter 2 for more details on this type of learning).

RoBERTa, which stands for robustly optimized BERT approach, is an improved version of BERT. It maintains the same overall architecture as BERT but employs several training and optimization improvements, such as larger batch sizes, more training data, and eliminating the next-sentence prediction task. These changes have resulted in RoBERTa achieving better performance on various natural language understanding tasks than BERT.

Decoders

Coming back to the original transformer architecture outlined in Figure 17-1, the multi-head self-attention mechanism in the decoder is similar to the one in the encoder, but it is masked to prevent the model from attending to future positions, ensuring that the predictions for position i can depend only on the known outputs at positions less than i. As illustrated in Figure 17-3, the decoder generates the output word by word.

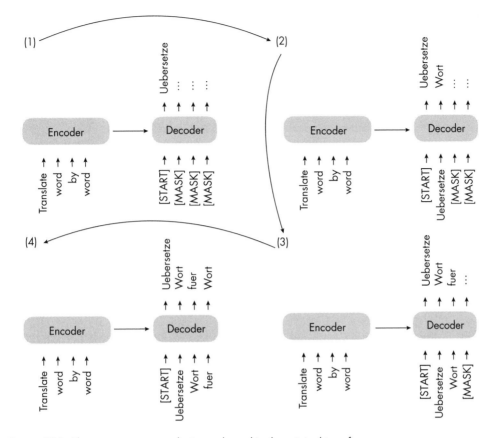

Figure 17-3: The next-sentence prediction task used in the original transformer

This masking (shown explicitly in Figure 17-3, although it occurs internally in the decoder's multi-head self-attention mechanism) is essential to maintaining the transformer model's autoregressive property during training and inference. This autoregressive property ensures that the model generates output tokens one at a time and uses previously generated tokens as context for generating the next word token.

Over the years, researchers have built upon the original encoder-decoder transformer architecture and developed several decoder-only models that have proven highly effective in various natural language processing tasks. The most notable models include the GPT family, which we briefly discussed in Chapter 14 and in various other chapters throughout the book.

GPT stands for *generative pretrained transformer*. The GPT series comprises decoder-only models pretrained on large-scale unsupervised text data and fine-tuned for specific tasks such as text classification, sentiment analysis, question answering, and summarization. The GPT models, including at the time of writing GPT-2, GPT-3, and GPT-4, have shown remarkable performance in various benchmarks and are currently the most popular architecture for natural language processing.

One of the most notable aspects of GPT models is their emergent properties. Emergent properties are the abilities and skills that a model develops due to its next-word prediction pretraining. Even though these models were taught only to predict the next word, the pretrained models are capable of text summarization, translation, question answering, classification, and more. Furthermore, these models can perform new tasks without updating the model parameters via in-context learning, which we'll discuss in more detail in Chapter 18.

Encoder-Decoder Hybrids

Next to the traditional encoder and decoder architectures, there have been advancements in the development of new encoder-decoder models that leverage the strengths of both components. These models often incorporate novel techniques, pretraining objectives, or architectural modifications to enhance their performance in various natural language processing tasks. Some notable examples of these new encoder-decoder models include BART and T5.

Encoder-decoder models are typically used for natural language processing tasks that involve understanding input sequences and generating output sequences, often with different lengths and structures. They are particularly good at tasks where there is a complex mapping between the input and output sequences and where it is crucial to capture the relationships between the elements in both sequences. Some common use cases for encoder-decoder models include text translation and summarization.

Terminology

All of these methods—encoder-only, decoder-only, and encoder-decoder models—are sequence-to-sequence models (often abbreviated as *seq2seq*). While we refer to BERT-style methods as "encoder-only," the description may be misleading since these methods also *decode* the embeddings into output tokens or text during pretraining. In other words, both encoder-only and decoder-only architectures perform decoding.

However, the encoder-only architectures, in contrast to decoder-only and encoder-decoder architectures, don't decode in an autoregressive fashion. *Autoregressive decoding* refers to generating output sequences one token at a time, conditioning each token on the previously generated tokens. Encoder-only models do not generate coherent output sequences in this manner. Instead, they focus on understanding the input text and producing task-specific outputs, such as labels or token predictions.

Contemporary Transformer Models

In brief, encoder-style models are popular for learning embeddings used in classification tasks, encoder-decoder models are used in generative tasks where the output heavily relies on the input (for example, translation and summarization), and decoder-only models are used for other types of generative tasks, including Q&A. Since the first transformer architecture emerged, hundreds of encoder-only, decoder-only, and encoder-decoder hybrids have been developed, as diagrammed in Figure 17-4.

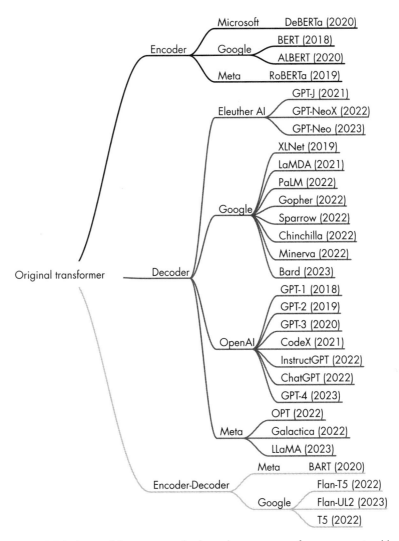

Figure 17-4: Some of the most popular large language transformers organized by architecture type and developer

While encoder-only models have gradually become less popular, decoder-only models like GPT have exploded in popularity, thanks to breakthroughs in text generation via GPT-3, ChatGPT, and GPT-4. However, encoder-only models are still very useful for training predictive models based on text embeddings as opposed to generating texts.

Exercises

17-1. As discussed in this chapter, BERT-style encoder models are pretrained using masked language modeling and next-sentence prediction pretraining objectives. How could we adopt such a pretrained model for a classification task (for example, predicting whether a text has a positive or negative sentiment)?

17-2. Can we fine-tune a decoder-only model like GPT for classification?

References

- The Bahdanau attention mechanism for RNNs: Dzmitry Bahdanau, Kyunghyun Cho, and Yoshua Bengio, "Neural Machine Translation by Jointly Learning to Align and Translate" (2014), *https://arxiv.org/abs/1409.0473*.

- The original BERT paper, which popularized encoder-style transformers with a masked word and a next-sentence prediction pretraining objective: Jacob Devlin et al., "BERT: Pre-training of Deep Bidirectional Transformers for Language Understanding" (2018), *https://arxiv.org/abs/1810.04805*.

- RoBERTa improves upon BERT by optimizing training procedures, using larger training datasets, and removing the next-sentence prediction task: Yinhan Liu et al., "RoBERTa: A Robustly Optimized BERT Pretraining Approach" (2019), *https://arxiv.org/abs/1907.11692*.

- The BART encoder-decoder architecture: Mike Lewis et al., "BART: Denoising Sequence-to-Sequence Pre-training for Natural Language Generation, Translation, and Comprehension" (2018), *https://arxiv.org/abs/1910.13461*.

- The T5 encoder-decoder architecture: Colin Raffel et al., "Exploring the Limits of Transfer Learning with a Unified Text-to-Text Transformer" (2019), *https://arxiv.org/abs/1910.10683*.

- The paper proposing the first GPT architecture: Alec Radford et al., "Improving Language Understanding by Generative Pre-Training" (2018), *https://cdn.openai.com/research-covers/language-unsupervised/language_understanding_paper.pdf*.

- The GPT-2 model: Alec Radford et al., "Language Models Are Unsupervised Multitask Learners" (2019), *https://www.semanticscholar.org/paper/Language-Models-are-Unsupervised-Multitask-Learners-Radford-Wu/9405cc0d6169988371b2755e573cc28650d14dfe*.

- The GPT-3 model: Tom B. Brown et al., "Language Models Are Few-Shot Learners" (2020), *https://arxiv.org/abs/2005.14165*.

18

USING AND FINE-TUNING PRETRAINED TRANSFORMERS

What are the different ways to use and fine-tune pretrained large language models?

The three most common ways to use and fine-tune pretrained LLMs include a feature-based approach, in-context prompting, and updating a subset of the model parameters. First, most pretrained LLMs or language transformers can be utilized without the need for further fine-tuning. For instance, we can employ a feature-based method to train a new downstream model, such as a linear classifier, using embeddings generated by a pretrained transformer. Second, we can showcase examples of a new task within the input itself, which means we can directly exhibit the expected outcomes without requiring any updates or learning from the model. This concept is also known as *prompting*. Finally, it's also possible to fine-tune all or just a small number of parameters to achieve the desired outcomes.

The following sections discuss these types of approaches in greater depth.

Using Transformers for Classification Tasks

Let's start with the conventional methods for utilizing pretrained transformers: training another model on feature embeddings, fine-tuning output layers, and fine-tuning all layers. We'll discuss these in the context of classification. (We will revisit prompting later in the section "In-Context Learning, Indexing, and Prompt Tuning" on page 116.)

In the feature-based approach, we load the pretrained model and keep it "frozen," meaning we do not update any parameters of the pretrained model. Instead, we treat the model as a feature extractor that we apply to our new dataset. We then train a downstream model on these embeddings. This downstream model can be any model we like (random forests, XGBoost, and so on), but linear classifiers typically perform best. This is likely because pretrained transformers like BERT and GPT already extract high-quality, informative features from the input data. These feature embeddings often capture complex relationships and patterns, making it easy for a linear classifier to effectively separate the data into different classes. Furthermore, linear classifiers, such as logistic regression machines and support vector machines, tend to have strong regularization properties. These regularization properties help prevent overfitting when working with high-dimensional feature spaces generated by pretrained transformers. This feature-based approach is the most efficient method since it doesn't require updating the transformer model at all. Finally, the embeddings can be precomputed for a given training dataset (since they don't change) when training a classifier for multiple training epochs.

Figure 18-1 illustrates how LLMs are typically created and adopted for downstream tasks using fine-tuning. Here, a pretrained model, trained on a general text corpus, is fine-tuned to perform tasks like German-to-English translation.

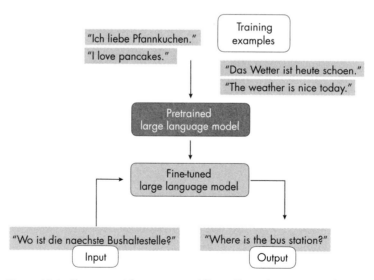

Figure 18-1: The general fine-tuning workflow of large language models

The conventional methods for fine-tuning pretrained LLMs include updating only the output layers, a method we'll refer to as *fine-tuning I*, and updating all layers, which we'll call *fine-tuning II*.

Fine-tuning I is similar to the feature-based approach described earlier, but it adds one or more output layers to the LLM itself. The backbone of the LLM remains frozen, and we update only the model parameters in these new layers. Since we don't need to backpropagate through the whole network, this approach is relatively efficient regarding throughput and memory requirements.

In fine-tuning II, we load the model and add one or more output layers, similarly to fine-tuning I. However, instead of backpropagating only through the last layers, we update *all* layers via backpropagation, making this the most expensive approach. While this method is computationally more expensive than the feature-based approach and fine-tuning I, it typically leads to better modeling or predictive performance. This is especially true for more specialized domain-specific datasets.

Figure 18-2 summarizes the three approaches described in this section so far.

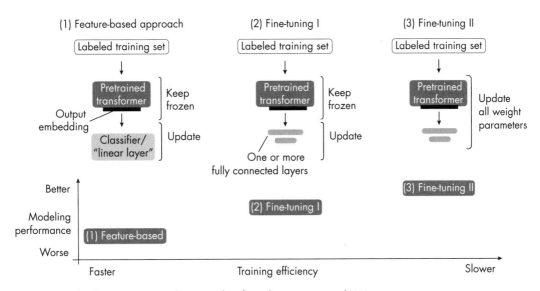

Figure 18-2: The three conventional approaches for utilizing pretrained LLMs

In addition to the conceptual summary of the three fine-tuning methods described in this section, Figure 18-2 also provides a rule-of-thumb guideline for these methods regarding training efficiency. Since fine-tuning II involves updating more layers and parameters than fine-tuning I, backpropagation is costlier for fine-tuning II. For similar reasons, fine-tuning II is costlier than a simpler feature-based approach.

In-Context Learning, Indexing, and Prompt Tuning

LLMs like GPT-2 and GPT-3 popularized the concept of *in-context learning*, often called *zero-shot* or *few-shot learning* in this context, which is illustrated in Figure 18-3.

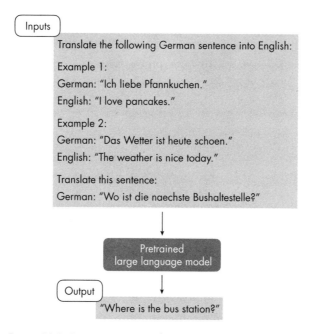

Figure 18-3: Prompting an LLM for in-context learning

As Figure 18-3 shows, in-context learning aims to provide context or examples of the task within the input or prompt, allowing the model to infer the desired behavior and generate appropriate responses. This approach takes advantage of the model's ability to learn from vast amounts of data during pretraining, which includes diverse tasks and contexts.

NOTE *The definition of few-shot learning, considered synonymous with in-context learning-based methods, differs from the conventional approach to few-shot learning discussed in Chapter 3.*

For example, suppose we want to use in-context learning for few-shot German–English translation using a large-scale pretrained language model like GPT-3. To do so, we provide a few examples of German–English translations to help the model understand the desired task, as follows:

```
Translate the following German sentences into English:

Example 1:
German: "Ich liebe Pfannkuchen."
English: "I love pancakes."

Example 2:
German: "Das Wetter ist heute schoen."
English: "The weather is nice today."

Translate this sentence:
German: "Wo ist die naechste Bushaltestelle?"
```

Generally, in-context learning does not perform as well as fine-tuning for certain tasks or specific datasets since it relies on the pretrained model's ability to generalize from its training data without further adapting its parameters for the particular task at hand.

However, in-context learning has its advantages. It can be particularly useful when labeled data for fine-tuning is limited or unavailable. It also enables rapid experimentation with different tasks without fine-tuning the model parameters in cases where we don't have direct access to the model or where we interact only with the model through a UI or API (for example, ChatGPT).

Related to in-context learning is the concept of *hard prompt tuning*, where *hard* refers to the non-differentiable nature of the input tokens. Where the previously described fine-tuning methods update the model parameters to better perform the task at hand, hard prompt tuning aims to optimize the prompt itself to achieve better performance. Prompt tuning does not modify the model parameters, but it may involve using a smaller labeled dataset to identify the best prompt formulation for the specific task. For example, to improve the prompts for the previous German–English translation task, we might try the following three prompting variations:

- `"Translate the German sentence '{german_sentence}' into English: {english_translation}"`
- `"German: '{german_sentence}' | English: {english_translation}"`
- `"From German to English: '{german_sentence}' -> {english_translation}"`

Prompt tuning is a resource-efficient alternative to parameter fine-tuning. However, its performance is usually not as good as full model fine-tuning, as it does not update the model's parameters for a specific task, potentially limiting its ability to adapt to task-specific nuances. Furthermore, prompt tuning can be labor intensive since it requires either human involvement comparing the quality of the different prompts or another similar method to do so. This is often known as *hard* prompting since, again, the input tokens are not differentiable. In addition, other methods exist that propose to use another LLM for automatic prompt generation and evaluation.

Yet another way to leverage a purely in-context learning-based approach is *indexing*, illustrated in Figure 18-4.

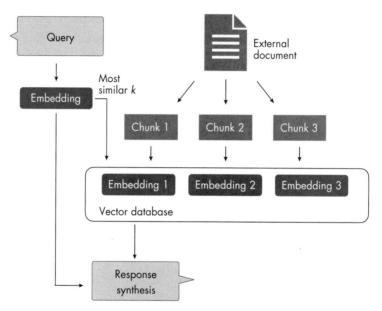

Figure 18-4: LLM indexing to retrieve information from external documents

In the context of LLMs, we can think of indexing as a workaround based on in-context learning that allows us to turn LLMs into information retrieval systems to extract information from external resources and websites. In Figure 18-4, an indexing module parses a document or website into smaller chunks, embedded into vectors that can be stored in a vector database. When a user submits a query, the indexing module computes the vector similarity between the embedded query and each vector stored in the database. Finally, the indexing module retrieves the top k most similar embeddings to synthesize the response.

Parameter-Efficient Fine-Tuning

In recent years, many methods have been developed to adapt pretrained transformers more efficiently for new target tasks. These methods are commonly referred to as *parameter-efficient fine-tuning*, with the most popular methods at the time of writing summarized in Figure 18-5.

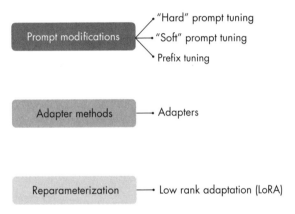

Figure 18-5: The main categories of parameter-efficient fine-tuning techniques, with popular examples

In contrast to the hard prompting approach discussed in the previous section, *soft prompting* strategies optimize embedded versions of the prompts. While in hard prompt tuning we modify the discrete input tokens, in soft prompt tuning we utilize trainable parameter tensors instead.

The idea behind soft prompt tuning is to prepend a trainable parameter tensor (the "soft prompt") to the embedded query tokens. The prepended tensor is then tuned to improve the modeling performance on a target dataset using gradient descent. In Python-like pseudocode, soft prompt tuning can be described as

```
x = EmbeddingLayer(input_ids)
x = concatenate([soft_prompt_tensor, x],
                dim=seq_len)
output = model(x)
```

where the soft_prompt_tensor has the same feature dimension as the embedded inputs produced by the embedding layer. Consequently, the modified input matrix has additional rows (as if it extended the original input sequence with additional tokens, making it longer).

Another popular prompt tuning method is prefix tuning. *Prefix tuning* is similar to soft prompt tuning, except that in prefix tuning, we prepend trainable tensors (soft prompts) to each transformer block instead of only the embedded inputs, which can stabilize the training. The implementation of prefix tuning is illustrated in the following pseudocode:

```
def transformer_block_with_prefix(x):
❶ soft_prompt = FullyConnectedLayers(# Prefix
     soft_prompt)                    # Prefix
❷ x = concatenate([soft_prompt, x],  # Prefix
                    dim=seq_len)      # Prefix
❸ residual = x
   x = SelfAttention(x)
   x = LayerNorm(x + residual)
   residual = x
   x = FullyConnectedLayers(x)
   x = LayerNorm(x + residual)
   return x
```

Listing 18-1: A transformer block modified for prefix tuning

Let's break Listing 18-1 into three main parts: implementing the soft prompt, concatenating the soft prompt (prefix) with the input, and implementing the rest of the transformer block.

First, the soft_prompt, a tensor, is processed through a set of fully connected layers ❶. Second, the transformed soft prompt is concatenated with the main input, x ❷. The dimension along which they are concatenated is denoted by seq_len, referring to the sequence length dimension. Third, the subsequent lines of code ❸ describe the standard operations in a transformer block, including self-attention, layer normalization, and feed-forward neural network layers, wrapped around residual connections.

As shown in Listing 18-1, prefix tuning modifies a transformer block by adding a trainable soft prompt. Figure 18-6 further illustrates the difference between a regular transformer block and a prefix tuning transformer block.

Regular transformer block

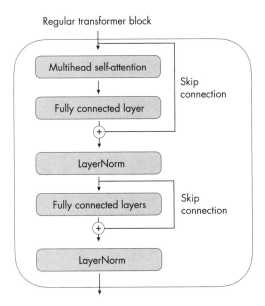

Transformer block with prefix

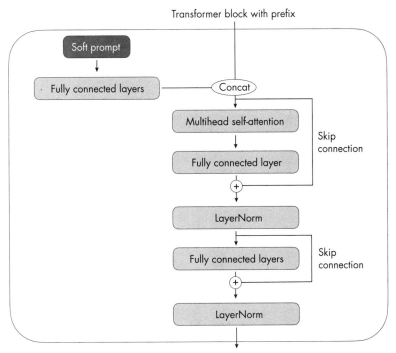

Figure 18-6: A regular transformer compared with prefix tuning

Both soft prompt tuning and prefix tuning are considered parameter efficient since they require training only the prepended parameter tensors and not the LLM parameters themselves.

Adapter methods are related to prefix tuning in that they add additional parameters to the transformer layers. In the original adapter method, additional fully connected layers were added after the multihead self-attention and existing fully connected layers in each transformer block, as illustrated in Figure 18-7.

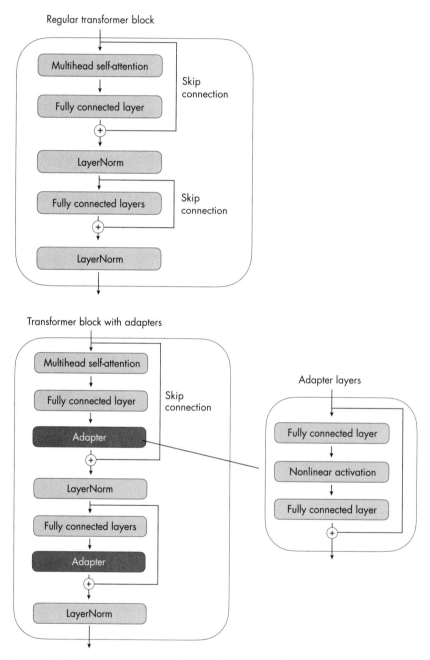

Figure 18-7: Comparison of a regular transformer block (left) and a transformer block with adapter layers

Only the new adapter layers are updated when training the LLM using the original adapter method, while the remaining transformer layers remain frozen. Since the adapter layers are usually small—the first fully connected layer in an adapter block projects its input into a low-dimensional representation, while the second layer projects it back into the original input dimension—this adapter method is usually considered parameter efficient.

In pseudocode, the original adapter method can be written as follows:

```
def transformer_block_with_adapter(x):
    residual = x
    x = SelfAttention(x)
    x = FullyConnectedLayers(x)  # Adapter
    x = LayerNorm(x + residual)
    residual = x
    x = FullyConnectedLayers(x)
    x = FullyConnectedLayers(x)  # Adapter
    x = LayerNorm(x + residual)
    return x
```

Low-rank adaptation (LoRA), another popular parameter-efficient fine-tuning method worth considering, refers to reparameterizing pretrained LLM weights using low-rank transformations. LoRA is related to the concept of *low-rank transformation*, a technique to approximate a high-dimensional matrix or dataset using a lower-dimensional representation. The lower-dimensional representation (or *low-rank approximation*) is achieved by finding a combination of fewer dimensions that can effectively capture most of the information in the original data. Popular low-rank transformation techniques include principal component analysis and singular vector decomposition.

For example, suppose ΔW represents the parameter update for a weight matrix of the LLM with dimension $\mathbb{R}^{A \times B}$. We can decompose the weight update matrix into two smaller matrices: $\Delta W = W_A W_B$, where $W_A \in \mathbb{R}^{A \times h}$ and $W_A \in \mathbb{R}^{h \times B}$. Here, we keep the original weight frozen and train only the new matrices W_A and W_B.

How is this method parameter efficient if we introduce new weight matrices? These new matrices can be very small. For example, if $A = 25$ and $B = 50$, then the size of ΔW is $25 \times 50 = 1{,}250$. If $h = 5$, then W_A has 125 parameters, W_B has 250 parameters, and the two matrices combined have only $125 + 250 = 375$ parameters in total.

After learning the weight update matrix, we can then write the matrix multiplication of a fully connected layer, as shown in this pseudocode:

```
def lora_forward_matmul(x):
    h = x . W  # Regular matrix multiplication
    h += x . (W_A . W_B) * scalar
    return h
```

Listing 18-2: Matrix multiplication with LoRA

In Listing 18-2, `scalar` is a scaling factor that adjusts the magnitude of the combined result (original model output plus low-rank adaptation). This balances the pretrained model's knowledge and the new task-specific adaptation.

According to the original paper introducing the LoRA method, models using LoRA perform slightly better than models using the adapter method across several task-specific benchmarks. Often, LoRA performs even better than models fine-tuned using the fine-tuning II method described earlier.

Reinforcement Learning with Human Feedback

The previous section focused on ways to make fine-tuning more efficient. Switching gears, how can we improve the modeling performance of LLMs via fine-tuning?

The conventional way to adapt or fine-tune an LLM for a new target domain or task is to use a supervised approach with labeled target data. For instance, the fine-tuning II approach allows us to adapt a pretrained LLM and fine-tune it on a target task such as sentiment classification, using a dataset that contains texts with sentiment labels like *positive*, *neutral*, and *negative*.

Supervised fine-tuning is a foundational step in training an LLM. An additional, more advanced step is *reinforcement learning with human feedback (RLHF)*, which can be used to further improve the model's alignment with human preferences. For example, ChatGPT and its predecessor, Instruct-GPT, are two popular examples of pretrained LLMs (GPT-3) fine-tuned using RLHF.

In RLHF, a pretrained model is fine-tuned using a combination of supervised learning and reinforcement learning. This approach was popularized by the original ChatGPT model, which was in turn based on Instruct-GPT. Human feedback is collected by having humans rank or rate different model outputs, providing a reward signal. The collected reward labels can be used to train a reward model that is then used to guide the LLMs' adaptation to human preferences. The reward model is learned via supervised learning, typically using a pretrained LLM as the base model, and is then used to adapt the pretrained LLM to human preferences via additional fine-tuning. The training in this additional fine-tuning stage uses a flavor of reinforcement learning called *proximal policy optimization*.

RLHF uses a reward model instead of training the pretrained model on the human feedback directly because involving humans in the learning process would create a bottleneck since we cannot obtain feedback in real time.

Adapting Pretrained Language Models

While fine-tuning all layers of a pretrained LLM remains the gold standard for adaption to new target tasks, several efficient alternatives exist for leveraging pretrained transformers. For instance, we can effectively apply LLMs to new tasks while minimizing computational costs and resources by utilizing feature-based methods, in-context learning, or parameter-efficient fine-tuning techniques.

The three conventional methods—feature-based approach, fine-tuning I, and fine-tuning II—provide different computational efficiency and performance trade-offs. Parameter-efficient fine-tuning methods like soft prompt tuning, prefix tuning, and adapter methods further optimize the adaptation process, reducing the number of parameters to be updated. Meanwhile, RLHF presents an alternative approach to supervised fine-tuning, potentially improving modeling performance.

In sum, the versatility and efficiency of pretrained LLMs continue to advance, offering new opportunities and strategies for effectively adapting these models to a wide array of tasks and domains. As research in this area progresses, we can expect further improvements and innovations in using pretrained language models.

Exercises

18-1. When does it make more sense to use in-context learning rather than fine-tuning, and vice versa?

18-2. In prefix tuning, adapters, and LoRA, how can we ensure that the model preserves (and does not forget) the original knowledge?

References

- The paper introducing the GPT-2 model: Alec Radford et al., "Language Models Are Unsupervised Multitask Learners" (2019), *https://www.semanticscholar.org/paper/Language-Models-are-Unsupervised-Multitask-Learners-Radford-Wu/9405cc0d6169988371b2755e573cc28650d14dfe*.

- The paper introducing the GPT-3 model: Tom B. Brown et al., "Language Models Are Few-Shot Learners" (2020), *https://arxiv.org/abs/2005.14165*.

- The automatic prompt engineering method, which proposes using another LLM for automatic prompt generation and evaluation: Yongchao Zhou et al., "Large Language Models Are Human-Level Prompt Engineers" (2023), *https://arxiv.org/abs/2211.01910*.

- LlamaIndex is an example of an indexing approach that leverages in-context learning: *https://github.com/jerryjliu/llama_index*.

- DSPy is a popular open source library for retrieval augmentation and indexing: *https://github.com/stanfordnlp/dsp*.

- A first instance of soft prompting: Brian Lester, Rami Al-Rfou, and Noah Constant, "The Power of Scale for Parameter-Efficient Prompt Tuning" (2021), *https://arxiv.org/abs/2104.08691*.

- The paper that first described prefix tuning: Xiang Lisa Li and Percy Liang, "Prefix-Tuning: Optimizing Continuous Prompts for Generation" (2021), *https://arxiv.org/abs/2101.00190*.

- The paper introducing the original adapter method: Neil Houlsby et al., "Parameter-Efficient Transfer Learning for NLP" (2019) *https://arxiv.org/abs/1902.00751*.

- The paper introducing the LoRA method: Edward J. Hu et al., "LoRA: Low-Rank Adaptation of Large Language Models" (2021), *https://arxiv.org/abs/2106.09685*.

- A survey of more than 40 research papers covering parameter-efficient fine-tuning methods: Vladislav Lialin, Vijeta Deshpande, and Anna Rumshisky, "Scaling Down to Scale Up: A Guide to Parameter-Efficient Fine-Tuning" (2023), *https://arxiv.org/abs/2303.15647*.

- The InstructGPT paper: Long Ouyang et al., "Training Language Models to Follow Instructions with Human Feedback" (2022), *https://arxiv.org/abs/2203.02155*.

- Proximal policy optimization, which is used for reinforcement learning with human feedback: John Schulman et al., "Proximal Policy Optimization Algorithms" (2017), *https://arxiv.org/abs/1707.06347*.

19

EVALUATING GENERATIVE LARGE LANGUAGE MODELS

What are the standard metrics for evaluating the quality of text generated by large language models, and why are these metrics useful?

Perplexity, BLEU, ROUGE, and BERTScore are some of the most common evaluation metrics used in natural language processing to assess the performance of LLMs across various tasks. Although there is ultimately no way around human quality judgments, human evaluations are tedious, expensive, hard to automate, and subjective. Hence, we develop metrics to provide objective summary scores to measure progress and compare different approaches.

This chapter discusses the difference between intrinsic and extrinsic performance metrics for evaluating LLMs, and then it dives deeper into popular metrics like BLEU, ROUGE, and BERTScore and provides simple hands-on examples for illustration purposes.

Evaluation Metrics for LLMs

The *perplexity metric* is directly related to the loss function used for pretraining LLMs and is commonly used to evaluate text generation and text completion models. Essentially, it quantifies the average uncertainty of the model in predicting the next word in a given context—the lower the perplexity, the better.

The *bilingual evaluation understudy (BLEU)* score is a widely used metric for evaluating the quality of machine-generated translations. It measures the overlap of n-grams between the machine-generated translation and a set of human-generated reference translations. A higher BLEU score indicates better performance, ranging from 0 (worst) to 1 (best).

The *recall-oriented understudy for gisting evaluation (ROUGE)* score is a metric primarily used for evaluating automatic summarization (and sometimes machine translation) models. It measures the overlap between the generated summary and reference summaries.

We can think of perplexity as an *intrinsic metric* and BLEU and ROUGE as *extrinsic metrics*. To illustrate the difference between the two types of metrics, think of optimizing the conventional cross entropy to train an image classifier. The cross entropy is a loss function we optimize during training, but our end goal is to maximize the classification accuracy. Since classification accuracy cannot be optimized directly during training, as it's not differentiable, we minimize the surrogate loss function like the cross entropy. Minimizing the cross entropy loss more or less correlates with maximizing the classification accuracy.

Perplexity is often used as an evaluation metric to compare the performance of different language models, but it is not the optimization target during training. BLEU and ROUGE are more related to classification accuracy, or rather precision and recall. In fact, BLEU is a precision-like score to evaluate the quality of a translated text, while ROUGE is a recall-like score to evaluate summarized texts.

The following sections discuss the mechanics of these metrics in more detail.

Perplexity

Perplexity is closely related to the cross entropy directly minimized during training, which is why we refer to perplexity as an *intrinsic metric*.

Perplexity is defined as $2^{H(p,\,q)/n}$, where $H(p, q)$ is the cross entropy between the true distribution of words p and the predicted distribution of words q, and n is the sentence length (the number of words or tokens) to normalize the score. As cross entropy decreases, perplexity decreases as well—the lower the perplexity, the better. While we typically compute the cross entropy using a natural logarithm, we calculate the cross entropy and perplexity with a base-2 logarithm for the intuitive relationship to hold. (However, whether we use a base-2 or natural logarithm is only a minor implementation detail.)

In practice, since the probability for each word in the target sentence is always 1, we compute the cross entropy as the logarithm of the probability scores returned by the model we want to evaluate. In other words, if we have the predicted probability score for each word in a sentence s, we can compute the perplexity directly as follows:

$$Perplexity(s) = 2^{-\frac{1}{n} \log_2(p(s))}$$

where s is the sentence or text we want to evaluate, such as "The quick brown fox jumps over the lazy dog," $p(s)$ is the probability scores returned by the model, and n is the number of words or tokens. For example, if the model returns the probability scores [0.99, 0.85, 0.89, 0.99, 0.99, 0.99, 0.99, 0.99], the perplexity is:

$$2^{-\frac{1}{8} \cdot \sum_i \log_2 p(w_i)}$$
$$= 2^{-\frac{1}{8} \cdot \sum \log_2(0.99 \times 0.85 \times 0.89 \times 0.99 \times 0.99 \times 0.99 \times 0.99 \times 0.99)}$$
$$= 1.043$$

If the sentence was "The fast black cat jumps over the lazy dog," with probabilities [0.99, 0.65, 0.13, 0.05, 0.21, 0.99, 0.99, 0.99], the corresponding perplexity would be 2.419.

You can find a code implementation and example of this calculation in the *supplementary/q19-evaluation-llms* subfolder at *https://github.com/rasbt/ MachineLearning-QandAI-book*.

BLEU Score

BLEU is the most popular and most widely used metric for evaluating translated texts. It's used in almost all LLMs capable of translation, including popular tools such as OpenAI's Whisper and GPT models.

BLEU is a reference-based metric that compares the model output to human-generated references and was first developed to capture or automate the essence of human evaluation. In short, BLEU measures the lexical overlap between the model output and the human-generated references based on a precision score.

In more detail, as a precision-based metric, BLEU counts how many words in the generated text (candidate text) occur in the reference text divided by the candidate text length (the number of words), where the reference text is a sample translation provided by a human, for example. This is commonly done for n-grams rather than individual words, but for simplicity, we will stick to words or 1-grams. (In practice, BLEU is often computed for 4-grams.)

Figure 19-1 demonstrates the BLEU score calculation, using the example of calculating the 1-gram BLEU score. The individual steps in Figure 19-1 illustrate how we compute the 1-gram BLEU score based on its individual components, the weighted precision times a brevity penalty. You can also find a code implementation of this calculation in the *supplementary/q15-text -augment* subfolder at *https://github.com/rasbt/MachineLearning-QandAI-book*.

(0) original = "Der schnelle braune Fuchs sprang ueber den faulen Hund" ← Sentence to translate

↓

(1) Count number of candidate words contained in the reference divided by the candidate length

reference =   " The quick brown fox jumped over the lazy dog "
candidate =   " The fast brown fox leaped over the dog "   ← Precision = 6/8 = 0.75

↓

(2) Problem: Maximum precision for repeated text

reference =   "The quick brown fox jumped over the lazy dog"
candidate =   "fox fox fox fox fox fox fox fox"   ← Precision = 8/8 = 1.0

↓

(3) Fix: Clip the count by the minimum number of times the word occurs in the reference and candidate:
  • Sum the clipped counts for all words in the candidate sentence
  • Sum the counts for all words in the candidate sentence
  • Calculate the weighted precision by dividing the total clipped count by the total count

reference =   "The quick brown fox jumped over the lazy dog"
candidate =   "fox fox fox fox fox fox fox fox"   ← WeightedPrecision = 1/8 = 0.13

reference =   " The quick brown fox jumped over the lazy dog "
candidate =   " The fast brown fox leaped over the dog "   ← WeightedPrecision = 6/8 = 0.75

↓

(4) Since for short translations it would be easier to score high, there is an additional brevity penalty:
$BrevityPenalty = \min\left(1, e^{1 - ReferenceLength/CandidateLength}\right) = 0.88$

reference =   " The quick brown fox jumped over the lazy dog "
candidate =   " The fast brown fox leaped over the dog "   ← BrevityPenalty × WeightedPrecision = 0.88 × 0.75 = 0.66

Figure 19-1: Calculating a 1-gram BLEU score

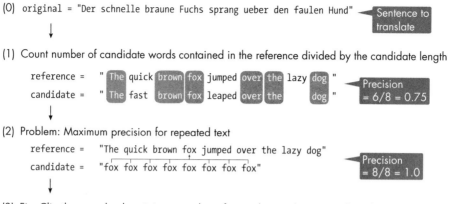

BLEU has several shortcomings, mostly owing to the fact that it measures string similarity, and similarity alone is not sufficient for capturing quality. For instance, sentences with similar words but different word orders might still score high, even though altering the word order can significantly change the meaning of a sentence and result in poor grammatical structure. Furthermore, since BLEU relies on exact string matches, it is sensitive to lexical variations and is incapable of identifying semantically similar translations that use synonyms or paraphrases. In other words, BLEU may assign lower scores to translations that are, in fact, accurate and meaningful.

The original BLEU paper found a high correlation with human evaluations, though this was disproven later.

Is BLEU flawed? Yes. Is it still useful? Also yes. BLEU is a helpful tool to measure or assess whether a model improves during training, as a proxy for fluency. However, it may not reliably give a correct assessment of the quality of the generated translations and is not well suited for detecting issues. In other words, it's best used as a model selection tool, not a model evaluation tool.

At the time of writing, the most popular alternatives to BLEU are METEOR and COMET (see the "References" section at the end of this chapter for more details).

ROUGE Score

While BLEU is commonly used for translation tasks, ROUGE is a popular metric for scoring text summaries.

There are many similarities between BLEU and ROUGE. The precision-based BLEU score checks how many words in the candidate translation occur in the reference translation. The ROUGE score also takes a flipped approach, checking how many words in the reference text appear in the generated text (here typically a summarization instead of a translation); this can be interpreted as a recall-based score.

Modern implementations compute ROUGE as an F1 score that is the harmonic mean of recall (how many words in the reference occur in the candidate text) and precision (how many words in the candidate text occur in the reference text). For example, Figure 19-2 shows a 1-gram ROUGE score computation (though in practice, ROUGE is often computed for bigrams, that is, 2-grams).

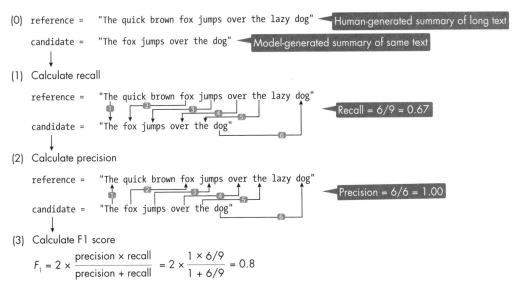

Figure 19-2: Computing ROUGE for 1-grams

There are other ROUGE variants beyond ROUGE-1 (the F1 score–based ROUGE score for 1-grams):

ROUGE-N   Measures the overlap of n-grams between the candidate and reference summaries. For example, ROUGE-1 would look at the overlap of individual words (1-grams), while ROUGE-2 would consider the overlap of 2-grams (bigrams).

ROUGE-L Measures the longest common subsequence (LCS) between the candidate and reference summaries. This metric captures the longest co-occurring in-order subsequence of words, which may have gaps in between them.

ROUGE-S Measures the overlap of *skip-bigrams*, or word pairs with a flexible number of words in between them. It can be useful to capture the similarity between sentences with different word orderings.

ROUGE shares similar weaknesses with BLEU. Like BLEU, ROUGE does not account for synonyms or paraphrases. It measures the n-gram overlap between the candidate and reference summaries, which can lead to lower scores for semantically similar but lexically different sentences. However, it's still worth knowing about ROUGE since, according to a study, *all* papers introducing new summarization models at computational linguistics conferences in 2021 used it, and 69 percent of those papers used *only* ROUGE.

BERTScore

Another more recently developed extrinsic metric is BERTScore.

For readers familiar with the inception score for generative vision models, BERTScore takes a similar approach, using embeddings from a pretrained model (for more on embeddings, see Chapter 1). Here, BERTScore measures the similarity between a candidate text and a reference text by leveraging the contextual embeddings produced by the BERT model (the encoder-style transformer discussed in Chapter 17).

The steps to compute BERTScore are as follows:

1. Obtain the candidate text via the LLM you want to evaluate (PaLM, LLaMA, GPT, BLOOM, and so on).

2. Tokenize the candidate and reference texts into subwords, preferably using the same tokenizer used for training BERT.

3. Use a pretrained BERT model to create the embeddings for all tokens in the candidate and reference texts.

4. Compare each token embedding in the candidate text to all token embeddings in the reference text, computing their cosine similarity.

5. Align each token in the candidate text with the token in the reference text that has the highest cosine similarity.

6. Compute the final BERTScore by taking the average similarity scores of all tokens in the candidate text.

Figure 19-3 further illustrates these six steps. You can also find a computational example in the *subfolder/q15-text-augment* subfolder at *https://github .com/rasbt/MachineLearning-QandAI-book*.

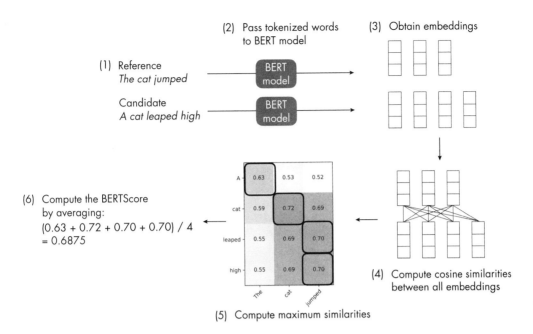

Figure 19-3: Computing the BERTScore step by step

BERTScore can be used for translations and summaries, and it captures the semantic similarity better than traditional metrics like BLEU and ROUGE. However, BERTScore is more robust in paraphrasing than BLEU and ROUGE and captures semantic similarity better due to its contextual embeddings. Also, it may be computationally more expensive than BLEU and ROUGE, as it requires using a pretrained BERT model for the evaluation. While BERTScore provides a useful automatic evaluation metric, it's not perfect and should be used alongside other evaluation techniques, including human judgment.

Surrogate Metrics

All metrics covered in this chapter are surrogates or proxies to evaluate how useful the model is in terms of measuring how well the model compares to human performance for accomplishing a goal. As mentioned earlier, the best way to evaluate LLMs is to assign human raters who judge the results. However, since this is often expensive and not easy to scale, we use the aforementioned metrics to estimate model performance. To quote from the InstructGPTpaper "Training Language Models to Follow Instructions with Human Feedback": "Public NLP datasets are not reflective of how our language models are used ... [They] are designed to capture tasks that are easy to evaluate with automatic metrics."

Besides perplexity, ROUGE, BLEU, and BERTScore, several other popular evaluation metrics are used to assess the predictive performance of LLMs.

Exercises

19-1. In step 5 of Figure 19-3, the cosine similarity between the two embeddings of "cat" is not 1.0, where 1.0 indicates a maximum cosine similarity. Why is that?

19-2. In practice, we might find that the BERTScore is not symmetric. This means that switching the candidate and reference sentences could result in different BERTScores for specific texts. How could we address this?

References

- The paper proposing the original BLEU method: Kishore Papineni et al., "BLEU: A Method for Automatic Evaluation of Machine Translation" (2002), *https://aclanthology.org/P02-1040/*.

- A follow-up study disproving BLEU's high correlation with human evaluations: Chris Callison-Burch, Miles Osborne, and Philipp Koehn, "Re-Evaluating the Role of BLEU in Machine Translation Research" (2006), *https://aclanthology.org/E06-1032/*.

- The shortcomings of BLEU, based on 37 studies published over 20 years: Benjamin Marie, "12 Critical Flaws of BLEU" (2022), *https://medium.com/@bnjmn_marie/12-critical-flaws-of-bleu-1d790ccbe1b1*.

- The paper proposing the original ROUGE method: Chin-Yew Lin, "ROUGE: A Package for Automatic Evaluation of Summaries" (2004), *https://aclanthology.org/W04-1013/*.

- A survey on the usage of ROUGE in conference papers: Sebastian Gehrmann, Elizabeth Clark, and Thibault Sellam, "Repairing the Cracked Foundation: A Survey of Obstacles in Evaluation Practices for Generated Text" (2022), *https://arxiv.org/abs/2202.06935*.

- BERTScore, an evaluation metric based on a large language model: Tianyi Zhang et al., "BERTScore: Evaluating Text Generation with BERT" (2019), *https://arxiv.org/abs/1904.09675*.

- A comprehensive survey on evaluation metrics for large language models: Asli Celikyilmaz, Elizabeth Clark, and Jianfeng Gao, "Evaluation of Text Generation: A Survey" (2021), *https://arxiv.org/abs/2006.14799*.

- METEOR is a machine translation metric that improves upon BLEU by using advanced matching techniques and aiming for better correlation with human judgment at the sentence level: Satanjeev Banerjee and Alon Lavie, "METEOR: An Automatic Metric for MT Evaluation with Improved Correlation with Human Judgments" (2005), *https://aclanthology.org/W05-0909/*.

- COMET is a neural framework that sets new standards for correlating machine translation quality with human judgments, using cross-lingual pretrained models and multiple types of evaluation: Ricardo Rei et al., "COMET: A Neural Framework for MT Evaluation" (2020), *https://arxiv.org/abs/2009.09025*.

- The InstructGPT paper: Long Ouyang et al., "Training Language Models to Follow Instructions with Human Feedback" (2022), *https://arxiv.org/abs/2203.02155*.

PART IV

PRODUCTION AND DEPLOYMENT

20

STATELESS AND STATEFUL TRAINING

What is the difference between stateless and stateful training workflows in the context of production and deployment systems?

Stateless training and stateful training refer to different ways of training a production model.

Stateless (Re)training

In stateless training, the more conventional approach, we first train an initial model on the original training set and then retrain it as new data arrives. Hence, stateless training is also commonly referred to as stateless *retraining*.

As Figure 20-1 shows, we can think of stateless retraining as a sliding window approach in which we retrain the initial model on different parts of the data from a given data stream.

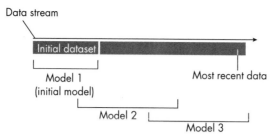

Figure 20-1: Stateless training replaces the model periodically.

For example, to update the initial model in Figure 20-1 (Model 1) to a newer model (Model 2), we train the model on 30 percent of the initial data and 70 percent of the most recent data at a given point in time.

Stateless retraining is a straightforward approach that allows us to adapt the model to the most recent changes in the data and feature-target relationships via retraining the model from scratch in user-defined checkpoint intervals. This approach is prevalent with conventional machine learning systems that cannot be fine-tuned as part of a transfer or self-supervised learning workflow (see Chapter 2). For example, standard implementations of tree-based models, such as random forests and gradient boosting (XGBoost, CatBoost, and LightGBM), fall into this category.

Stateful Training

In stateful training, we train the model on an initial batch of data and then update it periodically (as opposed to retraining it) when new data arrives.

As illustrated in Figure 20-2, we do not retrain the initial model (Model 1.0) from scratch; instead, we update or fine-tune it as new data arrives. This approach is particularly attractive for models compatible with transfer learning or self-supervised learning.

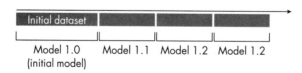

Figure 20-2: Stateful training updates models periodically.

The stateful approach mimics a transfer or self-supervised learning workflow where we adopt a pretrained model for fine-tuning. However, stateful training differs fundamentally from transfer and self-supervised learning because it updates the model to accommodate concept, feature, and label drifts. In contrast, transfer and self-supervised learning aim to adopt the model for a different classification task. For instance, in transfer learning, the target labels often differ. In self-supervised learning, we obtain the target labels from the dataset features.

One significant advantage of stateful training is that we do not need to store data for retraining; instead, we can use it to update the model as soon as it arrives. This is particularly attractive when data storage is a concern due to privacy or resource limitations.

Exercises

20-1. Suppose we train a classifier for stock trading recommendations using a random forest model, including the moving average of the stock price as a feature. Since new stock market data arrives daily, we are considering how to update the classifier daily to keep it up to date. Should we take a stateless training or stateless retraining approach to update the classifier?

20-2. Suppose we deploy a large language model (transformer) such as ChatGPT that can answer user queries. The dialogue interface includes thumbs-up and thumbs-down buttons so that users can give direct feedback based on the generated queries. While collecting the user feedback, we don't update the model immediately as new feedback arrives. However, we are planning to release a new or updated model at least once per month. Should we use stateless or stateful retraining for this model?

21

DATA-CENTRIC AI

What is data-centric AI, how does it compare to the conventional modeling paradigm, and how do we decide whether it's the right fit for a project?

Data-centric AI is a paradigm or workflow in which we keep the model training procedure fixed and iterate over the dataset to improve the predictive performance of a model. The following sections define what data-centric AI means in more detail and compare it to conventional model-centric approaches.

Data-Centric vs. Model-Centric AI

In the context of data-centric AI, we can think of the conventional workflow, which is often part of academic publishing, as model-centric AI. However, in an academic research setting, we are typically interested in developing new methods (for example, neural network architectures or loss functions). Here, we consider existing benchmark datasets to compare the new method to previous approaches and determine whether it is an improvement over the status quo.

Figure 21-1 summarizes the difference between data-centric and model-centric workflows.

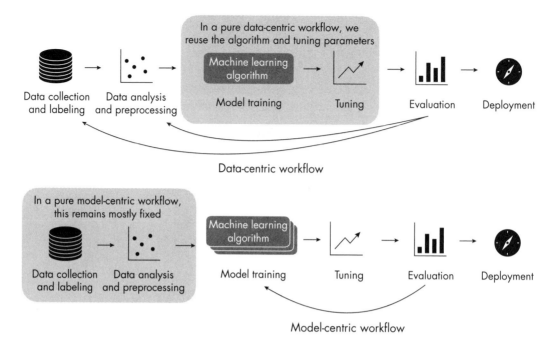

Figure 21-1: Data-centric versus model-centric machine learning workflow

While *data-centric AI* is a relatively new term, the idea behind it is not. Many people I've spoken with say they used a data-centric approach in their projects before the term was coined. In my opinion, data-centric AI was created to make "caring about data quality" attractive again, as data collection and curation are often considered tedious or thankless. This is analogous to how the term *deep learning* made neural networks interesting again in the early 2010s.

Do we need to choose between data-centric and model-centric AI, or can we rely on both? In short, data-centric AI focuses on changing the data to improve performance, while model-centric approaches focus on modifying the model to improve performance. Ideally, we should use both in an applied setting where we want to get the best possible predictive performance. However, in a research setting or an exploratory stage of an applied project, working with too many variables simultaneously is messy. If we change both model and data at once, it's hard to pinpoint which change is responsible for the improvement.

It is important to emphasize that data-centric AI is a paradigm and workflow, not a particular technique. Data-centric AI therefore implicitly includes the following:

- Analyses and modifications of training data, from outlier removal to missing data imputation
- Data synthesis and data augmentation techniques

- Data labeling and label-cleaning methods
- The classic active learning setting where a model suggests which data points to label

We consider an approach *data centric* if we change only the data (using the methods listed here), not the other aspects of the modeling pipeline.

In machine learning and AI, we often use the phrase "garbage in, garbage out," meaning that poor-quality data will result in a poor predictive model. In other words, we cannot expect a well-performing model from a low-quality dataset.

I've observed a common pattern in applied academic projects that attempt to use machine learning to replace an existing methodology. Often, researchers have only a small dataset of examples (say, hundreds of training examples). Labeling data is often expensive or considered boring and thus best avoided. In these cases, the researchers spend an unreasonable amount of time trying out different machine-learning algorithms and model tuning. To resolve this issue, investing additional time or resources in labeling additional data would be worthwhile.

The main advantage of data-centric AI is that it puts the data first so that if we invest resources to create a higher-quality dataset, all modeling approaches will benefit from it downstream.

Recommendations

Taking a data-centric approach is often a good idea in an applied project where we want to improve the predictive performance to solve a particular problem. In this context, it makes sense to start with a modeling baseline and improve the dataset since it's often more worthwhile than trying out bigger, more expensive models.

If our task is to develop a new or better methodology, such as a new neural network architecture or loss function, a model-centric approach might be a better choice. Using an established benchmark dataset without changing it makes it easier to compare the new modeling approach to previous work. Increasing the model size usually improves performance, but so does the addition of training examples. Assuming small training sets ($< 2k$) for classification, extractive question answering, and multiple-choice tasks, adding a hundred examples can result in the same performance gain as adding billions of parameters.

In a real-world project, alternating between data-centric and model-centric modes makes a lot of sense. Investing in data quality early on will benefit all models. Once a good dataset is available, we can begin to focus on model tuning to improve performance.

Exercises

21-1. A recent trend is the increased use of predictive analytics in healthcare. For example, suppose a healthcare provider develops an AI system that analyzes patients' electronic health records and provides recommendations for lifestyle changes or preventive measures. For this, the provider requires patients to monitor and share their health data (such as pulse and blood pressure) daily. Is this an example of data-centric AI?

21-2. Suppose we train a ResNet-34 convolutional neural network to classify images in the CIFAR-10 and ImageNet datasets. To reduce overfitting and improve classification accuracy, we experiment with data augmentation techniques such as image rotation and cropping. Is this approach data centric?

References

- An example of how adding more training data can benefit model performance more than an increase in model size: Yuval Kirstain et al., "A Few More Examples May Be Worth Billions of Parameters" (2021), *https://arxiv.org/abs/2110.04374*.

- Cleanlab is an open source library that includes methods for improving labeling errors and data quality in computer vision and natural language processing contexts: *https://github.com/cleanlab/cleanlab*.

22

SPEEDING UP INFERENCE

What are techniques to speed up model inference through optimization without changing the model architecture or sacrificing accuracy?

In machine learning and AI, *model inference* refers to making predictions or generating outputs using a trained model. The main general techniques for improving model performance during inference include parallelization, vectorization, loop tiling, operator fusion, and quantization, which are discussed in detail in the following sections.

Parallelization

One common way to achieve better parallelization during inference is to run the model on a batch of samples rather than on a single sample at a time. This is sometimes also referred to as *batched inference* and assumes that we are receiving multiple input samples or user inputs simultaneously or within a short time window, as illustrated in Figure 22-1.

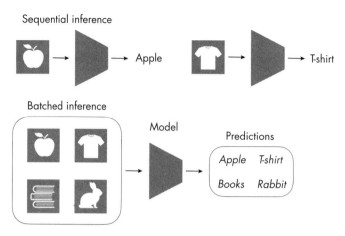

Figure 22-1: Sequential inference and batched inference

Figure 22-1 shows sequential inference processing one item at a time, which creates a bottleneck if there are several samples waiting to be classified. In batched inference, the model processes all four samples at the same time.

Vectorization

Vectorization refers to performing operations on entire data structures, such as arrays (tensors) or matrices, in a single step rather than using iterative constructs like for loops. Using vectorization, multiple operations from the loop are performed simultaneously using single instruction, multiple data (SIMD) processing, which is available on most modern CPUs.

This approach takes advantage of the low-level optimizations in many computing systems and often results in significant speedups. For example, it might rely on BLAS.

BLAS (which is short for *Basic Linear Algebra Subprograms*) is a specification that prescribes a set of low-level routines for performing common linear algebra operations such as vector addition, scalar multiplication, dot products, matrix multiplication, and others. Many array and deep learning libraries like NumPy and PyTorch use BLAS under the hood.

To illustrate vectorization with an example, suppose we wanted to compute the dot product between two vectors. The non-vectorized way of doing this would be to use a for loop, iterating over each element of the array one by one. However, this can be quite slow, especially for large arrays. With vectorization, you can perform the dot product operation on the entire array at once, as shown in Figure 22-2.

Classic for loop

```
x = [1.2, 2.2, 3.3, 4.4]
w = [5.5, 6.6, 7.7, 8.8]

output = 0.

for x_j, w_j in zip(x, w):
    output += x_j × w_j

print(output)
```

85.25

Vectorized implementation

```
import torch

x = torch.tensor([1.2, 2.2, 3.3, 4.4])
w = torch.tensor([5.5, 6.6, 7.7, 8.8])

x.dot(w)
```

tensor(85.2500)

Figure 22-2: A classic for loop versus a vectorized dot product computation in Python

In the context of linear algebra or deep learning frameworks like TensorFlow and PyTorch, vectorization is typically done automatically. This is because these frameworks are designed to work with multidimensional arrays (also known as *tensors*), and their operations are inherently vectorized. This means that when you perform functions using these frameworks, you automatically leverage the power of vectorization, resulting in faster and more efficient computations.

Loop Tiling

Loop tiling (also often referred to as *loop nest optimization*) is an advanced optimization technique to enhance data locality by breaking down a loop's iteration space into smaller chunks or "tiles." This ensures that once data is loaded into cache, all possible computations are performed on it before the cache is cleared.

Figure 22-3 illustrates the concept of loop tiling for accessing elements in a two-dimensional array. In a regular for loop, we iterate over columns and rows one element at a time, whereas in loop tiling, we subdivide the array into smaller tiles.

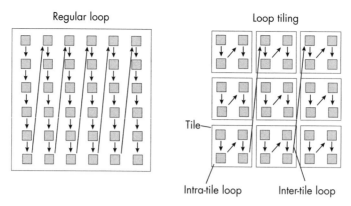

Figure 22-3: Loop tiling in a two-dimensional array

Note that in languages such as Python, we don't usually perform loop tiling, because Python and many other high-level languages do not allow control over cache memory like lower-level languages such as C and C++ do. These kinds of optimizations are often handled by underlying libraries like NumPy and PyTorch when performing operations on large arrays.

Operator Fusion

Operator fusion, sometimes called *loop fusion*, is an optimization technique that combines multiple loops into a single loop. This is illustrated in Figure 22-4, where two separate loops to calculate the sum and the product of an array of numbers are fused into a single loop.

```python
numbers = [1, 2, 3, 4, 5]

# First loop to calculate the sum
total_sum = 0
for num in numbers:
    total_sum += num

# Second loop to calculate the product
product = 1
for num in numbers:
    product *= num

print("Sum:", total_sum)
print("Product:", product)

Sum: 15
Product: 120
```

```python
numbers = [1, 2, 3, 4, 5]

# Single loop to calculate both
# the sum AND the product
total_sum = 0
product = 1
for num in numbers:
    total_sum += num
    product *= num

print("Sum:", total_sum)
print("Product:", product)

Sum: 15
Product: 120
```

Figure 22-4: Fusing two *for* loops (left) into one (right)

Operator fusion can improve the performance of a model by reducing the overhead of loop control, decreasing memory access times by improving cache performance, and possibly enabling further optimizations through vectorization.

You might think this behavior of vectorization would be incompatible with loop tiling, in which we break a for loop into multiple loops. However, these techniques are actually complementary, used for different optimizations, and applicable in different situations. Operator fusion is about reducing the total number of loop iterations and improving data locality when the entire data fits into cache. Loop tiling is about improving cache utilization when dealing with larger multidimensional arrays that do not fit into cache.

Related to operator fusion is the concept of *reparameterization*, which can often also be used to simplify multiple operations into one. Popular examples include training a network with multibranch architectures that are reparameterized into single-stream architectures during inference. This reparameterization approach differs from traditional operator fusion in that it does not merge multiple operations into a single operation. Instead, it rearranges the operations in the network to create a more efficient architecture for inference. In the so-called RepVGG architecture, for example, each branch during training consists of a series of convolutions. Once training is complete, the model is reparameterized into a single sequence of convolutions.

Quantization

Quantization reduces the computational and storage requirements of machine learning models, particularly deep neural networks. This technique involves converting the floating-point numbers (technically discrete but representing continuous values within a specific range) for implementing weights and biases in a trained neural network to more discrete, lower-precision representations such as integers. Using less precision reduces the model size and makes it quicker to execute, which can lead to significant improvements in speed and hardware efficiency during inference.

In the realm of deep learning, it has become increasingly common to quantize trained models down to 8-bit and 4-bit integers. These techniques are especially prevalent in the deployment of large language models.

There are two main categories of quantization. In *post-training quantization*, the model is first trained normally with full-precision weights, which are then quantized after training. *Quantization-aware training*, on the other hand, introduces the quantization step during the training process. This allows the model to learn to compensate for the effects of quantization, which can help maintain the model's accuracy.

However, it's important to note that quantization can occasionally lead to a reduction in model accuracy. Since this chapter focuses on techniques to speed up model inference *without* sacrificing accuracy, quantization is not as good a fit for this chapter as the previous categories.

NOTE *Other techniques to improve inference speeds include knowledge distillation and pruning, discussed in Chapter 6. However, these techniques affect the model architecture, resulting in smaller models, so they are out of scope for this chapter's question.*

Exercises

22-1. Chapter 7 covered several multi-GPU training paradigms to speed up model training. Using multiple GPUs can, in theory, also speed up model inference. However, in reality, this approach is often not the most efficient or most practical option. Why is that?

22-2. Vectorization and loop tiling are two strategies for optimizing operations that involve accessing array elements. What would be the ideal situation in which to use each?

References

- The official BLAS website: *https://www.netlib.org/blas/*.

- The paper that proposed loop tiling: Michael Wolfe, "More Iteration Space Tiling" (1989), *https://dl.acm.org/doi/abs/10.1145/76263 .76337*.

- RepVGG CNN architecture merging operations in inference mode: Xiaohan Ding et al., "RepVGG: Making VGG-style ConvNets Great Again" (2021), *https://arxiv.org/abs/2101.03697*.

- A new method for quantizing the weights in large language models down to 8-bit integer representations: Tim Dettmers et al., "LLM.int8(): 8-bit Matrix Multiplication for Transformers at Scale" (2022), *https://arxiv.org/abs/2208.07339*.

- A new method for quantizing the weights in LLMs farther down to 4-bit integers: Elias Frantar et al., "GPTQ: Accurate Post-Training Quantization for Generative Pre-trained Transformers" (2022), *https://arxiv.org/abs/2210.17323*.

23

DATA DISTRIBUTION SHIFTS

 What are the main types of data distribution shifts we may encounter after model deployment?

Data distribution shifts are one of the most common problems when putting machine learning and AI models into production. In short, they refer to the differences between the distribution of data on which a model was trained and the distribution of data it encounters in the real world. Often, these changes can lead to significant drops in model performance because the model's predictions are no longer accurate.

There are several types of distribution shifts, some of which are more problematic than others. The most common are covariate shift, concept drift, label shift, and domain shift; all discussed in more detail in the following sections.

Covariate Shift

Suppose $p(x)$ describes the distribution of the input data (for instance, the features), $p(y)$ refers to the distribution of the target variable (or class label distribution), and $p(y|x)$ is the distribution of the targets y given the inputs x.

Covariate shift happens when the distribution of the input data, $p(x)$, changes, but the conditional distribution of the output given the input, $p(y|x)$, remains the same.

Figure 23-1 illustrates covariate shift where both the feature values of the training data and the new data encountered during production follow a normal distribution. However, the mean of the new data has changed from the training data.

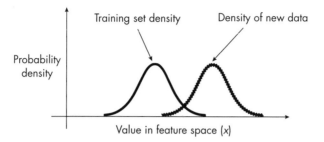

Figure 23-1: Training data and new data distributions differ under covariate shift.

For example, suppose we trained a model to predict whether an email is spam based on specific features. Now, after we embed the email spam filter in an email client, the email messages that customers receive have drastically different features. For example, the email messages are much longer and are sent from someone in a different time zone. However, if the way those features relate to an email being spam or not doesn't change, then we have a covariate shift.

Covariate shift is a very common challenge when deploying machine learning models. It means that the data the model receives in a live or production environment is different from the data on which it was trained. However, because the relationship between inputs and outputs, $p(y|x)$, remains the same under covariate shift, techniques are available to adjust for it.

A common technique to detect covariate shift is *adversarial validation*, which is covered in more detail in Chapter 29. Once covariate shift is detected, a common method to deal with it is *importance weighting*, which assigns different weights to the training example to emphasize or de-emphasize certain instances during training. Essentially, instances that are more likely to appear in the test distribution are given more weight, while instances that are less likely to occur are given less weight. This approach allows the model to focus more on the instances representative of the test data during training, making it more robust to covariate shift.

Label Shift

Label shift, sometimes referred to as *prior probability shift*, occurs when the class label distribution $p(y)$ changes, but the class-conditional distribution $p(y|x)$ remains unchanged. In other words, there is a significant change in the label distribution or target variable.

As an example of such a scenario, suppose we trained an email spam classifier on a balanced training dataset with 50 percent spam and 50 percent non-spam email. In contrast, in the real world, only 10 percent of email messages are spam.

A common way to address label shifts is to update the model using the weighted loss function, especially when we have an idea of the new distribution of the labels. This is essentially a form of importance weighting. By adjusting the weights in the loss function according to the new label distribution, we are incentivizing the model to pay more attention to certain classes that have become more common (or less common) in the new data. This helps align the model's predictions more closely with the current reality, improving its performance on the new data.

Concept Drift

Concept drift refers to the change in the mapping between the input features and the target variable. In other words, concept drift is typically associated with changes in the conditional distribution $p(y|x)$, such as the relationship between the inputs x and the output y.

Using the example of the spam email classifier from the previous section, the features of the email messages might remain the same, but *how* those features relate to whether an email is spam might change. This could be due to a new spamming strategy that wasn't present in the training data. Concept drift can be much harder to deal with than the other distribution shifts discussed so far since it requires continuous monitoring and potential model retraining.

Domain Shift

The terms *domain shift* and *concept drift* are used somewhat inconsistently across the literature and are sometimes taken to be interchangeable. In reality, the two are related but slightly different phenomena. *Concept drift* refers to a change in the function that maps from the inputs to the outputs, specifically to situations where the relationship between features and target variables changes as more data is collected over time.

In *domain shift*, the distribution of inputs, $p(x)$, and the conditional distribution of outputs given inputs, $p(y|x)$, both change. This is sometimes also called *joint distribution shift* due to the joint distribution $p(x$ and $y) = p(y|x) \cdot p(x)$. We can thus think of domain shift as a combination of both covariate shift and concept drift.

In addition, since we can obtain the marginal distribution $p(y)$ by integrating over the joint distribution $p(x, y)$ over the variable x (mathematically expressed as $p(y) = \int p(x, y)\, dx$), covariate drift and concept shift also imply label shift. (However, exceptions may exist where the change in $p(x)$ compensates for the change in $p(y|x)$ such that $p(y)$ may not change.) Conversely, label shift and concept drift usually also imply covariate shift.

To return once more to the example of email spam classification, domain shift would mean that the features (content and structure of email) *and* the relationship between the features and target both change over time. For instance, spam email in 2023 might have different features (new types of phishing schemes, new language, and so forth), and the definition of what constitutes spam might have changed as well. This type of shift would be the most challenging scenario for a spam filter trained on 2020 data, as it would have to adjust to changes in both the input data and the target concept.

Domain shift is perhaps the most difficult type of shift to handle, but monitoring model performance and data statistics over time can help detect domain shifts early. Once they are detected, mitigation strategies include collecting more labeled data from the target domain and retraining or adapting the model.

Types of Data Distribution Shifts

Figure 23-2 provides a visual summary of different types of data shifts in the context of a binary (2-class) classification problem, where the black circles refer to examples from one class and the diamonds refer to examples from another class.

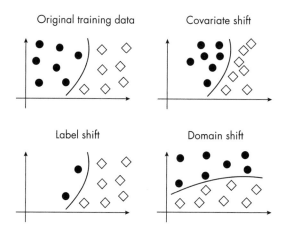

Figure 23-2: Different types of data shifts in a binary classification context

As noted in the previous sections, some types of distribution shift are more problematic than others. The least problematic among them is typically covariate shift. Here, the distribution of the input features, $p(x)$, changes between the training and testing data, but the conditional distribution of the output given the inputs, $p(y|x)$, remains constant. Since the underlying relationship between the inputs and outputs remains the same, the model trained on the training data can still apply, in principle, to the testing data and new data.

The most problematic type of distribution shift is typically joint distribution shift, where both the input distribution $p(x)$ and the conditional output distribution $p(y|x)$ change. This makes it particularly difficult for a model to adjust, as the learned relationship from the training data may no longer hold. The model has to cope with both new input patterns and new rules for making predictions based on those patterns.

However, the "severity" of a shift can vary widely depending on the real-world context. For example, even a covariate shift can be extremely problematic if the shift is severe or if the model cannot adapt to the new input distribution. On the other hand, a joint distribution shift might be manageable if the shift is relatively minor or if we have access to a sufficient amount of labeled data from the new distribution to retrain our model.

In general, it's crucial to monitor our models' performance and be aware of potential shifts in the data distribution so that we can take appropriate action if necessary.

Exercises

23-1. What is the big issue with importance weighting as a technique to mitigate covariate shift?

23-2. How can we detect these types of shifts in real-world scenarios, especially when we do not have access to labels for the new data?

References

- Recommendations and pointers to advanced mitigation techniques for avoiding domain shift: Abolfazl Farahani et al., "A Brief Review of Domain Adaptation" (2020), *https://arxiv.org/abs/2010.03978*.

PART V

PREDICTIVE PERFORMANCE AND MODEL EVALUATION

24

POISSON AND ORDINAL REGRESSION

When is it preferable to use Poisson regression over ordinal regression, and vice versa?

We usually use Poisson regression when the target variable represents count data (positive integers). As an example of count data, consider the number of colds contracted on an airplane or the number of guests visiting a restaurant on a given day. Besides the target variable representing counts, the data should also be Poisson distributed, which means that the mean and variance are roughly the same. (For large means, we can use a normal distribution to approximate a Poisson distribution.)

Ordinal data is a subcategory of categorical data where the categories have a natural order, such as 1 < 2 < 3, as illustrated in Figure 24-1. Ordinal data is often represented as positive integers and may look similar to count data. For example, consider the star rating on Amazon (1 star, 2 stars, 3 stars, and so on). However, ordinal regression does not make any assumptions about the distance between the ordered categories. Consider the following measure of disease severity: *severe* > *moderate* > *mild* > *none*. While we would typically map the disease severity variable to an integer representation (4 > 3 > 2 > 1), there is no assumption that the distance between 4 and 3 (severe and moderate) is the same as the distance between 2 and 1 (mild and none).

Count data with equal distances   Ordinal data with arbitrary distances

Figure 24-1: The distance between ordinal categories is arbitrary.

In short, we use Poisson regression for count data. We use ordinal regression when we know that certain outcomes are "higher" or "lower" than others, but we are not sure how much or if it even matters.

Exercises

24-1. Suppose we want to predict the number of goals a soccer player will score in a particular season. Should we solve this problem using ordinal regression or Poisson regression?

24-2. Suppose we ask someone to sort the last three movies they have watched based on their order of preference. Ignoring the fact that this dataset is a tad too small for machine learning, which approach would be best suited for this kind of data?

25

CONFIDENCE INTERVALS

What are the different ways to construct confidence intervals for machine learning classifiers?

There are several ways to construct confidence intervals for machine learning models, depending on the model type and the nature of your data. For instance, some methods are computationally expensive when working with deep neural networks and are thus more suitable to less resource-intensive machine learning models. Others require larger datasets to be reliable.

The following are the most common methods for constructing confidence intervals:

- Constructing normal approximation intervals based on a test set
- Bootstrapping training sets
- Bootstrapping the test set predictions
- Confidence intervals from retraining models with different random seeds

Before reviewing these in greater depth, let's briefly review the definition and interpretation of confidence intervals.

Defining Confidence Intervals

A *confidence interval* is a type of method to estimate an unknown population parameter. A *population parameter* is a specific measure of a statistical population, for example, a mean (average) value or proportion. By "specific" measure, I mean there is a single, exact value for that parameter for the entire population. Even though this value may not be known and often needs to be estimated from a sample, it is a fixed and definite characteristic of the population. A *statistical population*, in turn, is the complete set of items or individuals we study.

In a machine learning context, the population could be considered the entire possible set of instances or data points that the model may encounter, and the parameter we are often most interested in is the true generalization accuracy of our model on this population.

The accuracy we measure on the test set estimates the true generalization accuracy. However, it's subject to random error due to the specific sample of test instances we happened to use. This is where the concept of a confidence interval comes in. A 95 percent confidence interval for the generalization accuracy gives us a range in which we can be reasonably sure that the true generalization accuracy lies.

For instance, if we take 100 different data samples and compute a 95 percent confidence interval for each sample, approximately 95 of the 100 confidence intervals will contain the true population value (such as the generalization accuracy), as illustrated in Figure 25-1.

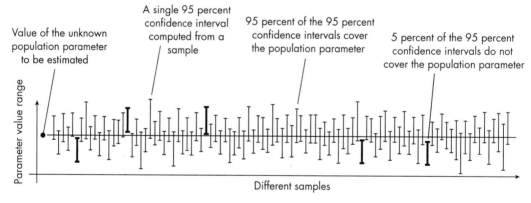

Figure 25-1: The concept of 95 percent confidence intervals

More concretely, if we were to draw 100 different representative test sets from the population (for instance, the entire possible set of instances that the model may encounter) and compute the 95 percent confidence interval for the generalization accuracy from each test set, we would expect about 95 of these intervals to contain the true generalization accuracy.

We can display confidence intervals in several ways. It is common to use a bar plot representation where the top of the bar represents the parameter value (for example, model accuracy) and the whiskers denote the upper and lower levels of the confidence interval (left chart of Figure 25-2). Alternatively, the confidence intervals can be shown without bars, as in the right chart of Figure 25-2.

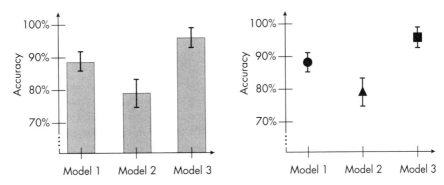

Figure 25-2: Two common plotting variants to illustrate confidence intervals

This visualization is functionally useful in a number of ways. For instance, when confidence intervals for two model performances do *not* overlap, it's a strong visual indicator that the performances are significantly different. Take the example of statistical significance tests, such as t-tests: if two 95 percent confidence intervals do not overlap, it strongly suggests that the difference between the two measurements is statistically significant at the 0.05 level.

On the other hand, if two 95 percent confidence intervals overlap, we cannot automatically conclude that there's no significant difference between the two measurements. Even when confidence intervals overlap, there can still be a statistically significant difference.

Alternatively, to provide more detailed information about the exact quantities, we can use a table view to express the confidence intervals. The two common notations are summarized in Table 25-1.

Table 25-1: Confidence Intervals

Model	Dataset A	Dataset B	Dataset C
1	89.1% ± 1.7%	. . .	. . .
2	79.5% ± 2.2%	. . .	. . .
3	95.2% ± 1.6%	. . .	. . .
Model	Dataset A	Dataset B	Dataset C
1	89.1% (87.4%, 90.8%)	. . .	. . .
2	79.5% (77.3%, 81.7%)	. . .	. . .
3	95.2% (93.6%, 96.8%)	. . .	. . .

The ± notation is often preferred if the confidence interval is *symmetric*, meaning the upper and lower endpoints are equidistant from the estimated parameter. Alternatively, the lower and upper confidence intervals can be written explicitly.

The Methods

The following sections describe the four most common methods of constructing confidence intervals.

Method 1: Normal Approximation Intervals

The normal approximation interval involves generating the confidence interval from a single train-test split. It is often considered the simplest and most traditional method for computing confidence intervals. This approach is especially appealing in the realm of deep learning, where training models is computationally costly. It's also desirable when we are interested in evaluating a specific model, instead of models trained on various data partitions like in k-fold cross-validation.

How does it work? In short, the formula for calculating the confidence interval for a predicted parameter (for example, the sample mean, denoted as $\bar{x}$), assuming a normal distribution, is expressed as $\bar{x} \pm z \times SE$.

In this formula, z represents the z-score, which indicates a particular value's number of standard deviations from the mean in a standard normal distribution. SE represents the standard error of the predicted parameter (in this case, the sample mean).

NOTE *Most readers will be familiar with z-score tables that are usually found in the back of introductory statistics textbooks. However, a more convenient and preferred way to obtain z-scores is to use functions like SciPy's* `stats.zscore` *function, which computes the z-scores for given confidence levels.*

For our scenario, the sample mean, denoted as $\bar{x}$, corresponds to the test set accuracy, ACC_{test}, a measure of successful predictions in the context of a binomial proportion confidence interval.

The standard error can be calculated under a normal approximation as follows:

$$SE = \sqrt{\frac{1}{n}\text{ACC}_{\text{test}}\left(1 - \text{ACC}_{\text{test}}\right)}$$

In this equation, n signifies the size of the test set. Substituting the standard error back into the previous formula, we obtain the following:

$$\text{ACC}_{\text{test}} \pm z\sqrt{\frac{1}{n}\text{ACC}_{\text{test}}\left(1 - \text{ACC}_{\text{test}}\right)}$$

Additional code examples to implement this method can also be found in the *supplementary/q25_confidence-intervals* subfolder in the supplementary code repository at *https://github.com/rasbt/MachineLearning-QandAI-book*.

While the normal approximation interval method is very popular due to its simplicity, it has some downsides. First, the normal approximation may not always be accurate, especially for small sample sizes or for data that is not normally distributed. In such cases, other methods of computing confidence intervals may be more accurate. Second, using a single train-test split does not provide information about the variability of the model performance across different splits of the data. This can be an issue if the performance is highly dependent on the specific split used, which may be the case if the dataset is small or if there is a high degree of variability in the data.

Method 2: Bootstrapping Training Sets

Confidence intervals serve as a tool for approximating unknown parameters. However, when we are restricted to just one estimate, such as the accuracy derived from a single test set, we must make certain assumptions to make this work. For example, when we used the normal approximation interval described in the previous section, we assumed normally distributed data, which may or may not hold.

In a perfect scenario, we would have more insight into our test set sample distribution. However, this would require access to many independent test datasets, which is typically not feasible. A workaround is the bootstrap method, which resamples existing data to estimate the sampling distribution.

NOTE   *In practice, when the test set is large enough, the normal distribution approximation will hold, thanks to the central limit theorem. This theorem states that the sum (or average) of a large number of independent, identically distributed random variables will approach a normal distribution, regardless of the underlying distribution of the individual variables. It is difficult to specify what constitutes a large-enough test set. However, under stronger assumptions than those of the central limit theorem, we can at least estimate the rate of convergence to the normal distribution using the Berry–Esseen theorem, which gives a more quantitative estimate of how quickly the convergence in the central limit theorem occurs.*

In a machine learning context, we can take the original dataset and draw a random sample *with replacement*. If the dataset has size n and we draw a random sample with replacement of size n, this implies that some data points will likely be duplicated in this new sample, whereas other data points are not sampled at all. We can then repeat this procedure for multiple rounds to obtain multiple training and test sets. This process is known as *out-of-bag bootstrapping*, illustrated in Figure 25-3.

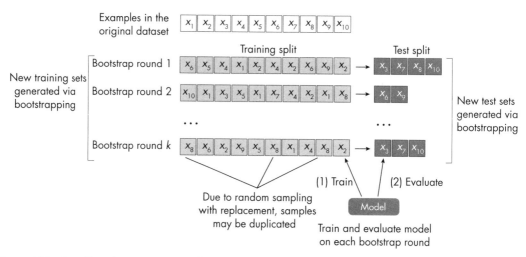

Figure 25-3: Out-of-bag bootstrapping evaluates models on resampled training sets.

Suppose we constructed k training and test sets. We can now take each of these splits to train and evaluate the model to obtain k test set accuracy estimates. Considering this distribution of test set accuracy estimates, we can take the range between the 2.5th and 97.5th percentile to obtain the 95 percent confidence interval, as illustrated in Figure 25-4.

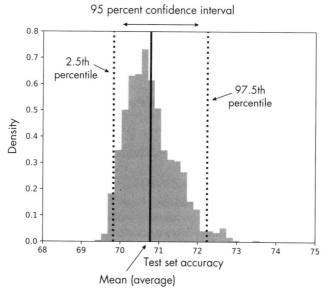

Figure 25-4: The distribution of test accuracies from 1,000 bootstrap samples, including a 95 percent confidence interval

Unlike the normal approximation interval method, we can consider this out-of-bag bootstrap approach to be more agnostic to the specific distribution. Ideally, if the assumptions for the normal approximation are satisfied, both methodologies would yield identical outcomes.

Since bootstrapping relies on resampling the existing test data, its downside is that it doesn't bring in any new information that could be available in a broader population or unseen data. Therefore, it may not always be able to generalize the performance of the model to new, unseen data.

Note that we are using the bootstrap sampling approach in this chapter instead of obtaining the train-test splits via k-fold cross-validation, because of the bootstrap's theoretical grounding via the central limit theorem discussed earlier. There are also more advanced out-of-bag bootstrap methods, such as the .632 and .632+ estimates, which are reweighting the accuracy estimates.

Method 3: Bootstrapping Test Set Predictions

An alternative approach to bootstrapping training sets is to bootstrap test sets. The idea is to train the model on the existing training set as usual and then to evaluate the model on bootstrapped test sets, as illustrated in Figure 25-5. After obtaining the test set performance estimates, we can then apply the percentile method described in the previous section.

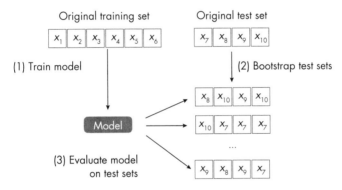

Figure 25-5: Bootstrapping the test set

Contrary to the prior bootstrap technique, this method uses a trained model and simply resamples the test set (instead of the training sets). This approach is especially appealing for evaluating deep neural networks, as it doesn't require retraining the model on the new data splits. However, a disadvantage of this approach is that it doesn't assess the model's variability toward small changes in the training data.

Method 4: Retraining Models with Different Random Seeds

In deep learning, models are commonly retrained using various random seeds since some random weight initializations may lead to much better models than others. How can we build a confidence interval from these experiments? If we assume that the sample means follow a normal distribution, we can employ a previously discussed method where we calculate the confidence interval around a sample mean, denoted as $\bar{x}$, as follows:

$$\bar{x} \pm z \times \text{SE}$$

Since in this context we often work with a relatively modest number of samples (for instance, models from 5 to 10 random seeds), assuming a t distribution is deemed more suitable than a normal distribution. Therefore, we substitute the z value with a t value in the preceding formula. (As the sample size increases, the t distribution tends to look more like the standard normal distribution, and the critical values [z and t] become increasingly similar.)

Furthermore, if we are interested in the average accuracy, denoted as $\overline{\text{ACC}}_{\text{test}}$, we consider $\text{ACC}_{\text{test},j}$ corresponding to a unique random seed j as a sample. The number of random seeds we evaluate would then constitute the sample size n. As such, we would calculate:

$$\overline{\text{ACC}}_{\text{test}} \pm t \times \text{SE}$$

Here, SE is the standard error, calculated as $\text{SE} = \text{SD}/\sqrt{n}$, while

$$\overline{\text{ACC}}_{\text{test}} = \frac{1}{r} \sum_{j=1}^{r} \text{ACC}_{\text{test},j}$$

is the average accuracy, which we compute over the r random seeds. The standard deviation SD is calculated as follows:

$$\text{SD} = \sqrt{\frac{\sum_j \left(ACC_{\text{test},j} - \overline{ACC}_{\text{test}}\right)^2}{r-1}}$$

To summarize, calculating the confidence intervals using various random seeds is another effective alternative. However, it is primarily beneficial for deep learning models. It proves to be costlier than both the normal approximation approach (method 1) and bootstrapping the test set (method 3), as it necessitates retraining the model. On the bright side, the outcomes derived from disparate random seeds provide us with a robust understanding of the model's stability.

Recommendations

Each possible method for constructing confidence intervals has its unique advantages and disadvantages. The normal approximation interval is cheap to compute but relies on the normality assumption about the distribution. The out-of-bag bootstrap is agnostic to these assumptions but is substantially more expensive to compute. A cheaper alternative is bootstrapping the test only, but this involves bootstrapping a smaller dataset and may be misleading for small or nonrepresentative test set sizes. Lastly, constructing confidence intervals from different random seeds is expensive but can give us additional insights into the model's stability.

Exercises

25-1. As mentioned earlier, the most common choice of confidence level is 95 percent confidence intervals. However, 90 percent and 99 percent are also common. Are 90 percent confidence intervals smaller or wider than 95 percent confidence intervals, and why is this the case?

25-2. In "Method 3: Bootstrapping Test Set Predictions" on page 169, we created test sets by bootstrapping and then applied the already trained model to compute the test set accuracy on each of these datasets. Can you think of a method or modification to obtain these test accuracies more efficiently?

References

- A detailed discussion of the pitfalls of concluding statistical significance from nonoverlapping confidence intervals: Martin Krzywinski and Naomi Altman, "Error Bars" (2013), *https://www.nature.com/articles/nmeth.2659*.

- A more detailed explanation of the binomial proportion confidence interval: *https://en.wikipedia.org/wiki/Binomial_proportion_confidence_interval*.

- For a detailed explanation of normal approximation intervals, see Section 1.7 of my article: "Model Evaluation, Model Selection, and Algorithm Selection in Machine Learning" (2018), *https://arxiv.org/abs/1811.12808*.

- Additional information on the central limit theorem for independent and identically distributed random variables: *https://en.wikipedia.org/wiki/Central_limit_theorem*.

- For more on the Berry–Esseen theorem: *https://en.wikipedia.org/wiki/Berry–Esseen_theorem*.

- The .632 bootstrap addresses a pessimistic bias of the regular out-of-bag bootstrapping approach: Bradley Efron, "Estimating the Error Rate of a Prediction Rule: Improvement on Cross-Validation" (1983), *https://www.jstor.org/stable/2288636*.

- The .632+ bootstrap corrects an optimistic bias introduced in the .632 bootstrap: Bradley Efron and Robert Tibshirani, "Improvements on Cross-Validation: The .632+ Bootstrap Method" (1997), *https://www.jstor.org/stable/2965703*.

- A deep learning research paper that discusses bootstrapping the test set predictions: Benjamin Sanchez-Lengeling et al., "Machine Learning for Scent: Learning Generalizable Perceptual Representations of Small Molecules" (2019), *https://arxiv.org/abs/1910.10685*.

26

CONFIDENCE INTERVALS VS. CONFORMAL PREDICTIONS

What are the differences between confidence intervals and conformal predictions, and when do we use one over the other?

Confidence intervals and conformal predictions are both statistical methods to estimate the range of plausible values for an unknown population parameter. As discussed in Chapter 25, a confidence interval quantifies the level of confidence that a population parameter lies within an interval. For instance, a 95 percent confidence interval for the mean of a population means that if we were to take many samples from the population and calculate the 95 percent confidence interval for each sample, we would expect the true population mean (average) to lie within these intervals 95 percent of the time. Chapter 25 covered several techniques for applying this method to estimate the prediction performance of machine learning models. Conformal predictions, on the other hand, are commonly used for creating prediction intervals, which are designed to cover a true outcome with a certain probability.

This chapter briefly explains what a prediction interval is and how it differs from confidence intervals, and then it explains how conformal predictions are, loosely speaking, a method for constructing prediction intervals.

Confidence Intervals and Prediction Intervals

Whereas a confidence interval focuses on parameters that characterize a population as a whole, a *prediction interval* provides a range of values for a single predicted target value. For example, consider the problem of predicting people's heights. Given a sample of 10,000 people from the population, we might conclude that the mean (average) height is 5 feet, 7 inches. We might also calculate a 95 percent confidence interval for this mean, ranging from 5 feet, 6 inches to 5 feet, 8 inches.

A *prediction interval*, however, is concerned with estimating not the height of the population but the height of an individual person. For example, given a weight of 185 pounds, a given person's prediction interval may fall between 5 feet 8 inches and 6 feet.

In a machine learning model context, we can use confidence intervals to estimate a population parameter such as the accuracy of a model (which refers to the performance on all possible prediction scenarios). In contrast, a prediction interval estimates the range of output values for a single given input example.

Prediction Intervals and Conformal Predictions

Both conformal predictions and prediction intervals are statistical techniques that estimate uncertainty for individual model predictions, but they do so in different ways and under different assumptions.

While prediction intervals often assume a particular data distribution and are tied to a specific type of model, conformal prediction methods are distribution free and can be applied to any machine learning algorithm.

In short, we can think of conformal predictions as a more flexible and generalizable form of prediction intervals. However, conformal predictions often require more computational resources than traditional methods for constructing prediction intervals, which involve resampling or permutation techniques.

Prediction Regions, Intervals, and Sets

In the context of conformal prediction, the terms *prediction interval*, *prediction set*, and *prediction region* are used to denote the plausible outputs for a given instance. The type of term used depends on the nature of the task.

In regression tasks where the output is a continuous variable, a *prediction interval* provides a range within which the true value is expected to fall with a certain level of confidence. For example, a model might predict that the price of a house is between $200,000 and $250,000.

In classification tasks, where the output is a discrete variable (the class labels), a *prediction set* includes all class labels that are considered plausible predictions for a given instance. For example, a model might predict that an image depicts either a cat, dog, or bird.

Prediction region is a more general term that can refer to either a prediction interval or a prediction set. It describes the set of outputs considered plausible by the model.

Computing Conformal Predictions

Now that we've introduced the difference between confidence intervals and prediction regions and learned how conformal prediction methods are related to prediction intervals, how exactly do conformal predictions work?

In short, conformal prediction methods provide a framework for creating prediction regions, sets of potential outcomes for a prediction task. Given the assumptions and methods used to construct them, these regions are designed to contain the true outcome with a certain probability.

For classifiers, a prediction region for a given input is a set of labels such that the set contains the true label with a given confidence (typically 95 percent), as illustrated in Figure 26-1.

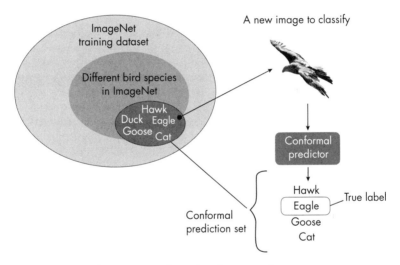

Figure 26-1: Prediction regions for a classification task

As depicted in Figure 26-1, the ImageNet dataset consists of a subset of bird species. Some bird species in ImageNet belong to one of the following classes: *hawk*, *duck*, *eagle*, or *goose*. ImageNet also contains other animals, for example, cats. For a new image to classify (here, an eagle), the conformal prediction set consists of classes such that the true label, *eagle*, is contained within this set with 95 percent probability. Often, this includes closely related classes, such as *hawk* and *goose* in this case. However, the prediction set can also include less closely related class labels, such as *cat*.

To sketch the concept of computing prediction regions step by step, let's suppose we train a machine learning classifier for images. Before the model is trained, the dataset is typically split into three parts: a training set, a calibration set, and a test set. We use the training set to train the model and the calibration set to obtain the parameters for the conformal prediction regions. We can then use the test set to assess the performance of the conformal predictor. A typical split ratio might be 60 percent training data, 20 percent calibration data, and 20 percent test data.

The first step after training the model on the training set is to define a *nonconformity measure*, a function that assigns a numeric score to each instance in the calibration set based on how "unusual" it is. This could be based on the distance to the classifier's decision boundary or, more commonly, 1 minus the predicted probability of a class label. The higher the score is, the more unusual the instance is.

Before using conformal predictions for new data points, we use the nonconformity scores from the calibration set to compute a quantile threshold. This threshold is a probability level such that, for example, 95 percent of the instances in the calibration set (if we choose a 95 percent confidence level) have nonconformity scores below this threshold. This threshold is then used to determine the prediction regions for new instances, ensuring that the predictions are calibrated to the desired confidence level.

Once we have the threshold value, we can compute prediction regions for new data. Here, for each possible class label (each possible output of your classifier) for a given instance, we check whether its nonconformity score is below the threshold. If it is, then we include it in the prediction set for that instance.

A Conformal Prediction Example

Let's illustrate this process of making conformal predictions with an example using a simple conformal prediction method known as the *score method*. Suppose we train a classifier on a training set to distinguish between three species of birds: sparrows, robins, and hawks. Suppose the predicted probabilities for a calibration dataset are as follows:

Sparrow [0.95, 0.9, 0.85, 0.8, 0.75]

Robin [0.7, 0.65, 0.6, 0.55, 0.5]

Hawk [0.4, 0.35, 0.3, 0.25, 0.2]

As depicted here, we have a calibration set consisting of 15 examples, five for each of the three classes. Note that a classifier returns three probability scores for each training example: one probability corresponding to each of the three classes (*Sparrow*, *Robin*, and *Hawk*). Here, however, we've selected only the probability for the true class label. For example, we may obtain the values [0.95, 0.02, 0.03] for the first calibration example with the true label *Sparrow*. In this case, we kept only 0.95.

Next, after we obtain the previous probability scores, we can compute the nonconformity score as 1 minus the probability, as follows:

Sparrow   [0.05, 0.1, 0.15, 0.2, 0.25]

Robin   [0.3, 0.35, 0.4, 0.45, 0.5]

Hawk   [0.6, 0.65, 0.7, 0.75, 0.8]

Considering a confidence level of 0.95, we now select a threshold such that 95 percent of these nonconformity scores fall below that threshold. Based on the nonconformity scores in this example, this threshold is 0.8. We can then use this threshold to construct the prediction sets for new instances we want to classify.

Now suppose we have a new instance (a new image of a bird) that we want to classify. We calculate the nonconformity score of this new bird image, assuming it belongs to each bird species (class label) in the training set:

Sparrow   0.26

Robin   0.45

Hawk   0.9

In this case, the *Sparrow* and *Robin* nonconformity scores fall below the threshold of 0.8. Thus, the prediction set for this input is [*Sparrow, Robin*]. In other words, this tells us that, on average, the true class label is included in the prediction set 95 percent of the time.

A hands-on code example implementing the score method can be found in the *supplementary/q26_conformal-prediction* subfolder at *https://github.com/ rasbt/MachineLearning-QandAI-book*.

The Benefits of Conformal Predictions

In contrast to using class-membership probabilities returned from classifiers, the major benefits of conformal prediction are its theoretical guarantees and its generality. Conformal prediction methods don't make any strong assumptions about the distribution of the data or the model being used, and they can be applied in conjunction with any existing machine learning algorithm to provide confidence measures for predictions.

Confidence intervals have asymptotic coverage guarantees, which means that the coverage guarantee holds in the limit as the sample (test set) size goes to infinity. This doesn't necessarily mean that confidence intervals work for only very large sample sizes, but rather that their properties are more firmly guaranteed as the sample size increases. Confidence intervals therefore rely on asymptotic properties, meaning that their guarantees become more robust as the sample size grows.

In contrast, conformal predictions provide finite-sample guarantees, ensuring that the coverage probability is achieved for any sample size. For example, if we specify a 95 percent confidence level for a conformal prediction method and generate 100 calibration sets with corresponding prediction sets, the method will include the true class label for 95 out of the 100 test points. This holds regardless of the size of the calibration sets.

While conformal prediction has many advantages, it does not always provide the tightest possible prediction intervals. Sometimes, if the underlying assumptions of a specific classifier hold, that classifier's own probability estimates might offer tighter and more informative intervals.

Recommendations

A confidence interval tells us about our level of uncertainty about the model's properties, such as the prediction accuracy of a classifier. A prediction interval or conformal prediction output tells us about the level of uncertainty in a specific prediction from the model. Both are very important in understanding the reliability and performance of our model, but they provide different types of information.

For example, a confidence interval for the prediction accuracy of a model can be helpful for comparing and evaluating models and for deciding which model to deploy. On the other hand, a prediction interval can be helpful for using a model in practice and understanding its predictions. For instance, it can help identify cases where the model is unsure and may need additional data, human oversight, or a different approach.

Exercises

26-1. Prediction set sizes can vary between instances. For example, we may encounter a prediction set size of 1 for a given instance and for another, a set size of 3. What does the prediction set size tell us?

26-2. Chapters 25 and 26 focused on classification methods. Could we use conformal prediction and confidence intervals for regression too?

References

- MAPIE is a popular library for conformal predictions in Python: *https://mapie.readthedocs.io/*.

- For more on the score method used in this chapter: Christoph Molnar, *Introduction to Conformal Prediction with Python* (2023), *https://christophmolnar.com/books/conformal-prediction/*.

- In addition to the score method, several other variants of conformal prediction methods exist. For a comprehensive collection of conformal prediction literature and resources, see the Awesome Conformal Prediction page: *https://github.com/valeman/awesome-conformal-prediction*.

27

PROPER METRICS

What are the three properties of a distance function that make it a *proper* metric?

Metrics are foundational to mathematics, computer science, and various other scientific domains. Understanding the fundamental properties that define a good distance function to measure distances or differences between points or datasets is important. For instance, when dealing with functions like loss functions in neural networks, understanding whether they behave like proper metrics can be instrumental in knowing how optimization algorithms will converge to a solution.

This chapter analyzes two commonly utilized loss functions, the mean squared error and the cross-entropy loss, to demonstrate whether they meet the criteria for proper metrics.

The Criteria

To illustrate the criteria of a proper metric, consider two vectors or points **v** and **w** and their distance $d(\mathbf{v}, \mathbf{w})$, as shown in Figure 27-1.

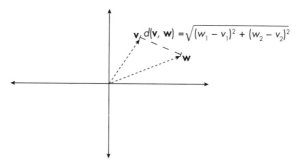

Figure 27-1: The Euclidean distance between two 2D vectors

The criteria of a proper metric are the following:

- The distance between two points is always non-negative, $d(\mathbf{v}, \mathbf{w}) \geq 0$, and can be 0 only if the two points are identical, that is, $\mathbf{v} = \mathbf{w}$.

- The distance is symmetric; for instance, $d(\mathbf{v}, \mathbf{w}) = d(\mathbf{w}, \mathbf{v})$.

- The distance function satisfies the *triangle inequality* for any three points: $\mathbf{v}, \mathbf{w}, \mathbf{x}$, meaning $d(\mathbf{v}, \mathbf{w}) \leq d(\mathbf{v}, \mathbf{x}) + d(\mathbf{x}, \mathbf{w})$.

To better understand the triangle inequality, think of the points as vertices of a triangle. If we consider any triangle, the sum of two of the sides is always larger than the third side, as illustrated in Figure 27-2.

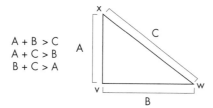

Figure 27-2: Triangle inequality

Consider what would happen if the triangle inequality depicted in Figure 27-2 weren't true. If the sum of the lengths of sides AB and BC was shorter than AC, then sides AB and BC would not meet to form a triangle; instead, they would fall short of each other. Thus, the fact that they meet and form a triangle demonstrates the triangle inequality.

The Mean Squared Error

The *mean squared error (MSE)* loss computes the squared Euclidean distance between a target variable y and a predicted target value $\hat{y}$:

$$\text{MSE} = \frac{1}{n} \sum_{i=1}^{n} \left(y^{(i)} - \hat{y}^{(i)} \right)^2$$

The index i denotes the ith data point in the dataset or sample. Is this loss function a proper metric?

For simplicity's sake, we will consider the *squared error (SE)* loss between two data points (though the following insights also hold for the MSE). As shown in the following equation, the SE loss quantifies the squared difference between the predicted and actual values for a single data point, while the MSE loss averages these squared differences over all data points in a dataset:

$$SE(y, \hat{y}) = (y - \hat{y})^2$$

In this case, the SE satisfies the first part of the first criterion: the distance between two points is always non-negative. Since we are raising the difference to the power of 2, it cannot be negative.

How about the second criterion, that the distance can be 0 only if the two points are identical? Due to the subtraction in the SE, it is intuitive to see that it can be 0 only if the prediction matches the target variable, $y = \hat{y}$. As with the first criterion, we can use the square to confirm that SE satisfies the second criterion: we have $(y - \hat{y})^2 = (\hat{y} - y)^2$.

At first glance, it seems that the squared error loss also satisfies the third criterion, the triangle inequality. Intuitively, you can check this by choosing three arbitrary numbers, here 1, 2, 3:

- $(1 - 2)^2 \leq (1 - 3)^2 + (2 - 3)^2$
- $(1 - 3)^2 \leq (1 - 2)^2 + (2 - 3)^2$
- $(2 - 3)^2 \leq (1 - 2)^2 + (1 - 3)^2$

However, there are values for which this is not true. For example, consider the values $a = 0$, $b = 2$, and $c = 1$. This gives us $d(a, b) = 4$, $d(a, c) = 1$, and $d(b, c) = 1$, such that we have the following scenario, which violates the triangle inequality:

- $(0 - 2)^2 \nleq (0 - 1)^2 + (2 - 1)^2$
- $(2 - 1)^2 \leq (0 - 1)^2 + (0 - 2)^2$
- $(0 - 1)^2 \leq (0 - 2)^2 + (1 - 2)^2$

Since it does not satisfy the triangle inequality via the example above, we conclude that the (mean) squared error loss is not a proper metric.

However, if we change the squared error into the *root-squared error*

$$\sqrt{(y - \hat{y})^2}$$

the triangle inequality can be satisfied:

$$\sqrt{(0 - 2)^2} \leq \sqrt{(0 - 1)^2} + \sqrt{(2 - 1)^2}$$

NOTE *You might be familiar with the L_2 distance or Euclidean distance, which is known to satisfy the triangle inequality. These two distance metrics are equivalent to the root-squared error when considering two scalar values.*

The Cross-Entropy Loss

Cross entropy is used to measure the distance between two probability distributions. In machine learning contexts, we use the discrete cross-entropy loss (CE) between class label y and the predicted probability p when we train logistic regression or neural network classifiers on a dataset consisting of n training examples:

$$\text{CE}(\mathbf{y}, \mathbf{p}) = -\frac{1}{n} \sum_{i=1}^{n} y^{(i)} \times \log \left(p^{(i)} \right)$$

Is this loss function a proper metric? Again, for simplicity's sake, we will look at the cross-entropy function (H) between only two data points:

$$H(y, p) = -y \times \log(p)$$

The cross-entropy loss satisfies one part of the first criterion: the distance is always non-negative because the probability score is a number in the range [0, 1]. Hence, $\log(p)$ ranges between $-\infty$ and 0. The important part is that the H function includes a negative sign. Hence, the cross entropy ranges between ∞ and 0 and thus satisfies one aspect of the first criterion shown above.

However, the cross-entropy loss is not 0 for two identical points. For example, $H(0.9, 0.9) = -0.9 \times \log(0.9) = 0.095$.

The second criterion shown above is also violated by the cross-entropy loss because the loss is not symmetric: $-y \times \log(p) \neq -p \times \log(y)$. Let's illustrate this with a concrete, numeric example:

- If $y = 1$ and $p = 0.5$, then $-1 \times \log(0.5) = 0.693$.
- If $y = 0.5$ and $p = 1$, then $-0.5 \times \log(1) = 0$.

Finally, the cross-entropy loss does not satisfy the triangle inequality, $H(r, p) \geq H(r, q) + H(q, p)$. Let's illustrate this with an example as well. Suppose we choose $r = 0.9$, $p = 0.5$, and $q = 0.4$. We have:

- $H(0.9, 0.5) = 0.624$
- $H(0.9, 0.4) = 0.825$
- $H(0.4, 0.5) = 0.277$

As you can see, $0.624 \geq 0.825 + 0.277$ does not hold here.

In conclusion, while the cross-entropy loss is a useful loss function for training neural networks via (stochastic) gradient descent, it is not a proper distance metric, as it does not satisfy any of the three criteria.

Exercises

27-1. Suppose we consider using the mean absolute error (MAE) as an alternative to the root mean square error (RMSE) for measuring the performance of a machine learning model, where MAE = $\frac{1}{n} \sum_{i=1}^{n} |y^{(i)} - \hat{y}^{(i)}|$ and RMSE = $\sqrt{\frac{1}{n} \sum_{i=1}^{n} (y^{(i)} - \hat{y}^{(i)})^2}$. However, a colleague argues that the MAE is not a proper distance metric in metric space because it involves an absolute value, so we should use the RMSE instead. Is this argument correct?

27-2. Based on your answer to the previous question, would you say that the MAE is better or is worse than the RMSE?

28

THE K IN K-FOLD CROSS-VALIDATION

k-fold cross-validation is a common choice for evaluating machine learning classifiers because it lets us use all training data to simulate how well a machine learning algorithm might perform on new data. What are the advantages and disadvantages of choosing a large *k*?

We can think of *k*-fold cross-validation as a workaround for model evaluation when we have limited data. In machine learning model evaluation, we care about the generalization performance of our model, that is, how well it performs on new data. In *k*-fold cross-validation, we use the training data for model selection and evaluation by partitioning it into *k* validation rounds and folds. If we have *k* folds, we have *k* iterations, leading to *k* different models, as illustrated in Figure 28-1.

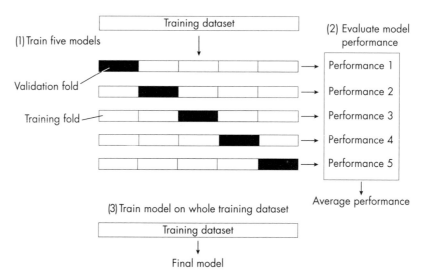

Figure 28-1: An example of k-fold cross-validation for model evaluation where k = 5

Using k-fold cross-validation, we usually evaluate the performance of a particular hyperparameter configuration by computing the average performance over the k models. This performance reflects or approximates the performance of a model trained on the complete training dataset after evaluation.

The following sections cover the trade-offs of selecting values for k in k-fold cross-validation and address the challenges of large k values and their computational demands, especially in deep learning contexts. We then discuss the core purposes of k and how to choose an appropriate value based on specific modeling needs.

Trade-offs in Selecting Values for k

If k is too large, the training sets are too similar between the different rounds of cross-validation. The k models are thus very similar to the model we obtain by training on the whole training set. In this case, we can still leverage the advantage of k-fold cross-validation: evaluating the performance for the entire training set via the held-out validation fold in each round. (Here, we obtain the training set by concatenating all $k - 1$ training folds in a given iteration.) However, a disadvantage of a large k is that it is more challenging to analyze how the machine learning algorithm with the particular choice of hyperparameter setting behaves on different training datasets.

Besides the issue of too-similar datasets, running k-fold cross-validation with a large value of k is also computationally more demanding. A larger k is more expensive since it increases both the number of iterations and the training set size at each iteration. This is especially problematic if we work with relatively large models that are expensive to train, such as contemporary deep neural networks.

A common choice for k is typically 5 or 10, for practical and historical reasons. A study by Ron Kohavi (see "References" at the end of this chapter) found that $k = 10$ offers a good bias and variance trade-off for classical machine learning algorithms, such as decision trees and naive Bayes classifiers, on a handful of small datasets.

For example, in 10-fold cross-validation, we use 9/10 (90 percent) of the data for training in each round, whereas in 5-fold cross-validation, we use only 4/5 (80 percent) of the data, as shown in Figure 28-2.

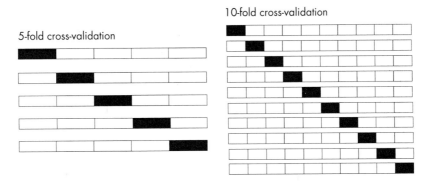

Figure 28-2: A comparison of 5-fold and 10-fold cross-validation

However, this does not mean large training sets are bad, since they can reduce the pessimistic bias of the performance estimate (mostly a good thing) if we assume that the model training can benefit from more training data. (See Figure 5-1 on page 24 for an example of a learning curve.)

In practice, both a very small and a very large k may increase variance. For instance, a larger k makes the training folds more similar to each other since a smaller proportion is left for the held-out validation sets. Since the training folds are more similar, the models in each round will be more similar. In practice, we may observe that the variance of the held-out validation fold scores is more similar for larger values of k. On the other hand, when k is large, the validation sets are small, so they may contain more random noise or be more susceptible to quirks of the data, leading to more variation in the validation scores across the different folds. Even though the models themselves are more similar (since the training sets are more similar), the validation scores may be more sensitive to the particularities of the small validation sets, leading to higher variance in the overall cross-validation score.

Determining Appropriate Values for k

When deciding upon an appropriate value of k, we are often guided by computational performance and conventions. However, it's worthwhile to define the purpose and context of using k-fold cross-validation. For example, if we care primarily about approximating the predictive performance of the final model, using a large k makes sense. This way, the training folds are very similar to the combined training dataset, yet we still get to evaluate the model on all data points via the validation folds.

On the other hand, if we care to evaluate how sensitive a given hyperparameter configuration and training pipeline is to different training datasets, then choosing a smaller number for k makes more sense.

Since most practical scenarios consist of two steps—tuning hyperparameters and evaluating the performance of a model—we can also consider a two-step procedure. For instance, we can use a smaller k during hyperparameter tuning. This will help speed up the hyperparameter search and probe the hyperparameter configurations for robustness (in addition to the average performance, we can also consider the variance as a selection criterion). Then, after hyperparameter tuning and selection, we can increase the value of k to evaluate the model.

However, reusing the same dataset for model selection and evaluation introduces biases, and it is usually better to use a separate test set for model evaluation. Also, nested cross-validation may be preferred as an alternative to k-fold cross-validation.

Exercises

28-1. Suppose we want to provide a model with as much training data as possible. We consider using *leave-one-out cross-validation (LOOCV)*, a special case of k-fold cross-validation where k is equal to the number of training examples, such that the validation folds contain only a single data point. A colleague mentions that LOOCV is defective for discontinuous loss functions and performance measures such as classification accuracy. For instance, for a validation fold consisting of only one example, the accuracy is always either 0 (0 percent) or 1 (99 percent). Is this really a problem?

28-2. This chapter discussed model selection and model evaluation as two use cases of k-fold cross-validation. Can you think of other use cases?

References

- For a longer and more detailed explanation of why and how to use k-fold cross-validation, see my article: "Model Evaluation, Model Selection, and Algorithm Selection in Machine Learning" (2018), *https://arxiv.org/abs/1811.12808*.

- The paper that popularized the recommendation of choosing $k = 5$ and $k = 10$: Ron Kohavi, "A Study of Cross-Validation and Bootstrap for Accuracy Estimation and Model Selection" (1995), *https://dl.acm .org/doi/10.5555/1643031.1643047*.

29

TRAINING AND TEST SET DISCORDANCE

Suppose we train a model that performs much better on the test dataset than on the training dataset. Since a similar model configuration previously worked well on a similar dataset, we suspect something might be unusual with the data. What are some approaches for looking into training and test set discrepancies, and what strategies can we use to mitigate these issues?

Before investigating the datasets in more detail, we should check for technical issues in the data loading and evaluation code. For instance, a simple sanity check is to temporarily replace the test set with the training set and to reevaluate the model. In this case, we should see identical training and test set performances (since these datasets are now identical). If we notice a discrepancy, we likely have a bug in the code; in my experience, such bugs are frequently related to incorrect shuffling or inconsistent (often missing) data normalization.

If the test set performance is much better than the training set performance, we can rule out overfitting. More likely, there are substantial differences in the training and test data distributions. These distributional differences may affect both the features and the targets. Here, plotting the target or label distributions of training and test data is a good idea. For example, a common issue is that the test set is missing certain class labels if the dataset was not shuffled properly before splitting it into training and test data. For small tabular datasets, it is also feasible to compare feature distributions in the training and test sets using histograms.

Looking at feature distributions is a good approach for tabular data, but this is trickier for image and text data. A relatively easy and more general approach to check for discrepancies between training and test sets is adversarial validation.

Adversarial validation, illustrated in Figure 29-1, is a technique to identify the degree of similarity between the training and test data. We first merge the training and test sets into a single dataset, and then we create a binary target variable that distinguishes between training and test data. For instance, we can use a new *Is test?* label where we assign the label 0 to training data and the label 1 to test data. We then use k-fold cross-validation or repartition the dataset into a training set and a test set and train a machine learning model as usual. Ideally, we want the model to perform poorly, indicating that the training and test data distributions are similar. On the other hand, if the model performs well in predicting the *Is test?* label, it suggests a discrepancy between the training and test data that we need to investigate further.

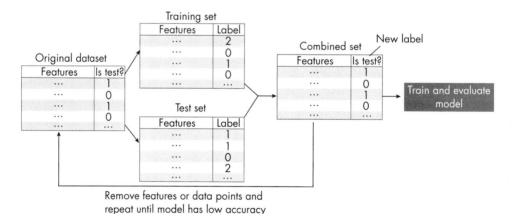

Figure 29-1: The adversarial validation workflow for detecting training and test set discrepancies

What mitigation techniques should we use if we detect a training–test set discrepancy using adversarial validation? If we're working with a tabular dataset, we can remove features one at a time to see if this helps address the issue, as spurious features can sometimes be highly correlated with the target variable. To implement this strategy, we can use sequential feature selection algorithms with an updated objective. For example, instead of maximizing classification accuracy, we can minimize classification accuracy. For cases where removing features is not so trivial (such as with image and text data), we can also investigate whether removing individual training instances that are different from the test set can address the discrepancy issue.

Exercises

29-1. What is a good performance baseline for the adversarial prediction task?

29-2. Since training datasets are often bigger than test datasets, adversarial validation often results in an imbalanced prediction problem (with a majority of examples labeled as *Is test?* being false instead of true). Is this an issue, and if so, how can we mitigate that?

30

LIMITED LABELED DATA

Suppose we plot a learning curve (as shown in Figure 5-1 on page 24, for example) and find the machine learning model overfits and could benefit from more training data. What are some different approaches for dealing with limited labeled data in supervised machine learning settings?

In lieu of collecting more data, there are several methods related to regular supervised learning that we can use to improve model performance in limited labeled data regimes.

Improving Model Performance with Limited Labeled Data

The following sections explore various machine learning paradigms that help in scenarios where training data is limited.

Labeling More Data

Collecting additional training examples is often the best way to improve the performance of a model (a learning curve is a good diagnostic for this). However, this is often not feasible in practice, because acquiring high-quality data can be costly, computational resources and storage might be insufficient, or the data may be hard to access.

Bootstrapping the Data

Similar to the techniques for reducing overfitting discussed in Chapter 5, it can be helpful to "bootstrap" the data by generating modified (augmented) or artificial (synthetic) training examples to boost the performance of the predictive model. Of course, improving the quality of data can also lead to the improved predictive performance of a model, as discussed in Chapter 21.

Transfer Learning

Transfer learning describes training a model on a general dataset (for example, ImageNet) and then fine-tuning the pretrained target dataset (for example, a dataset consisting of different bird species), as outlined in Figure 30-1.

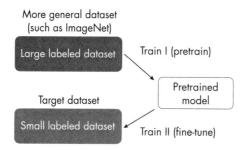

Figure 30-1: The process of transfer learning

Transfer learning is usually done in the context of deep learning, where model weights can be updated. This is in contrast to tree-based methods, since most decision tree algorithms are nonparametric models that do not support iterative training or parameter updates.

Self-Supervised Learning

Similar to transfer learning, in self-supervised learning, the model is pretrained on a different task before being fine-tuned to a target task for which only limited data exists. However, self-supervised learning usually relies on label information that can be directly and automatically extracted from unlabeled data. Hence, self-supervised learning is also often called *unsupervised pretraining*.

Common examples of self-supervised learning include the *next word* (used in GPT, for example) or *masked word* (used in BERT, for example) pretraining tasks in language modeling, covered in more detail in Chapter 17. Another intuitive example from computer vision includes *inpainting*: predicting the missing part of an image that was randomly removed, illustrated in Figure 30-2.

Figure 30-2: Inpainting for self-supervised learning

For more detail on self-supervised learning, see Chapter 2.

Active Learning

In active learning, illustrated in Figure 30-3, we typically involve manual labelers or users for feedback during the learning process. However, instead of labeling the entire dataset up front, active learning includes a prioritization scheme for suggesting unlabeled data points for labeling to maximize the machine learning model's performance.

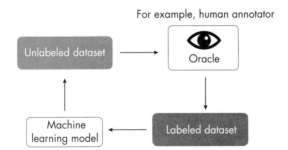

Figure 30-3: In active learning, a model queries an oracle for labels.

The term *active learning* refers to the fact that the model actively selects data for labeling. For example, the simplest form of active learning selects data points with high prediction uncertainty for labeling by a human annotator (also referred to as an *oracle*).

Few-Shot Learning

In a few-shot learning scenario, we often deal with extremely small datasets that include only a handful of examples per class. In research contexts, 1-shot (one example per class) and 5-shot (five examples per class) learning scenarios are very common. An extreme case of few-shot learning is zero-shot learning, where no labels are provided. Popular examples of zero-shot learning include GPT-3 and related language models, where the user has to provide all the necessary information via the input prompt, as illustrated in Figure 30-4.

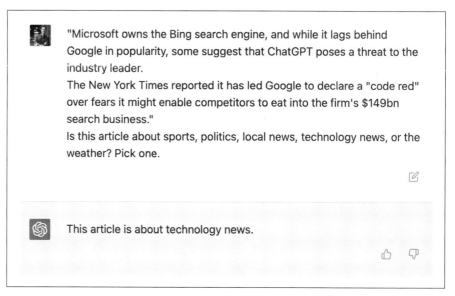

Figure 30-4: Zero-shot classification with ChatGPT

For more detail on few-shot learning, see Chapter 3.

Meta-Learning

Meta-learning involves developing methods that determine how machine learning algorithms can best learn from data. We can therefore think of meta-learning as "learning to learn." The machine learning community has developed several approaches for meta-learning. Within the machine learning community, the term *meta-learning* doesn't just represent multiple subcategories and approaches; it is also occasionally employed to describe related yet distinct processes, leading to nuances in its interpretation and application.

Meta-learning is one of the main subcategories of few-shot learning. Here, the focus is on learning a good feature extraction module, which converts support and query images into vector representations. These vector representations are optimized for determining the predicted class of the query example via comparisons with the training examples in the support set. (This form of meta-learning is illustrated in Chapter 3 on page 17.)

Another branch of meta-learning unrelated to the few-shot learning approach is focused on extracting metadata (also called *meta-features*) from datasets for supervised learning tasks, as illustrated in Figure 30-5. The meta-features are descriptions of the dataset itself. For example, these can include the number of features and statistics of the different features (kurtosis, range, mean, and so on).

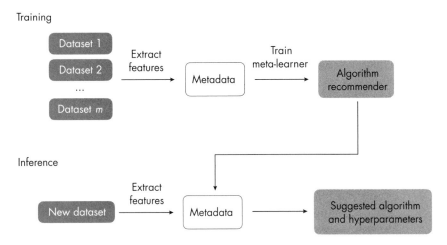

Figure 30-5: The meta-learning process involving the extraction of metadata

The extracted meta-features provide information for selecting a machine learning algorithm for the dataset at hand. Using this approach, we can narrow down the algorithm and hyperparameter search spaces, which helps reduce overfitting when the dataset is small.

Weakly Supervised Learning

Weakly supervised learning, illustrated in Figure 30-6, involves using an external label source to generate labels for an unlabeled dataset. Often, the labels created by a weakly supervised labeling function are more noisy or inaccurate than those produced by a human or domain expert, hence the term *weakly* supervised. We can develop or adopt a rule-based classifier to create the labels in weakly supervised learning; these rules usually cover only a subset of the unlabeled dataset.

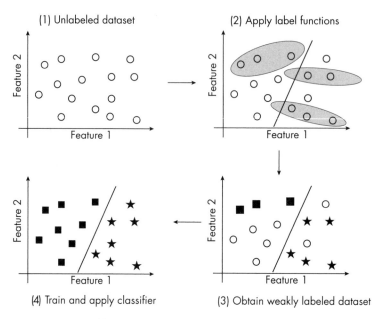

Figure 30-6: *Weakly supervised learning uses external labeling functions to train machine learning models.*

Let's return to the example of email spam classification from Chapter 23 to illustrate a rule-based approach for data labeling. In weak supervision, we could design a rule-based classifier based on the keyword *SALE* in the email subject header line to identify a subset of spam emails. Note that while we may use this rule to label certain emails as spam positive, we should not apply this rule to label emails without *SALE* as non-spam. Instead, we should either leave those unlabeled or apply a different rule to them.

There is a subcategory of weakly supervised learning referred to as PU-learning. In *PU-learning*, which is short for *positive-unlabeled learning*, we label and learn only from positive examples.

Semi-Supervised Learning

Semi-supervised learning is closely related to weakly supervised learning: it also involves creating labels for unlabeled instances in the dataset. The main difference between these two methods lies in *how* we create the labels. In weak supervision, we create labels using an external labeling function that is often noisy, inaccurate, or covers only a subset of the data. In semi-supervision, we do not use an external label function; instead, we leverage the structure of the data itself. We can, for example, label additional data points based on the density of neighboring labeled data points, as illustrated in Figure 30-7.

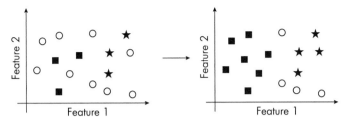

Figure 30-7: Semi-supervised learning

While we can apply weak supervision to an entirely unlabeled dataset, semi-supervised learning requires at least a portion of the data to be labeled. In practice, it is possible first to apply weak supervision to label a subset of the data and then to use semi-supervised learning to label instances that were not captured by the labeling functions.

Thanks to their close relationship, semi-supervised learning is sometimes referred to as a subcategory of weakly supervised learning, and vice versa.

Self-Training

Self-training falls somewhere between semi-supervised learning and weakly supervised learning. For this technique, we train a model to label the dataset or adopt an existing model to do the same. This model is also referred to as a *pseudo-labeler*.

Self-training does not guarantee accurate labels and is thus related to weakly supervised learning. Moreover, while we use or adopt a machine learning model for this pseudo-labeling, self-training is also related to semi-supervised learning.

An example of self-training is knowledge distillation, discussed in Chapter 6.

Multi-Task Learning

Multi-task learning trains neural networks on multiple, ideally related tasks. For example, if we are training a classifier to detect spam emails, spam classification is the main task. In multi-task learning, we can add one or more related tasks for the model to solve, referred to as *auxiliary tasks*. For the spam email example, an auxiliary task could be classifying the email's topic or language.

Typically, multi-task learning is implemented via multiple loss functions that have to be optimized simultaneously, with one loss function for each task. The auxiliary tasks serve as an inductive bias, guiding the model to prioritize hypotheses that can explain multiple tasks. This approach often results in models that perform better on unseen data.

There are two subcategories of multi-task learning: multi-task learning with hard parameter sharing and multi-task learning with soft parameter sharing. Figure 30-8 illustrates the difference between these two methods.

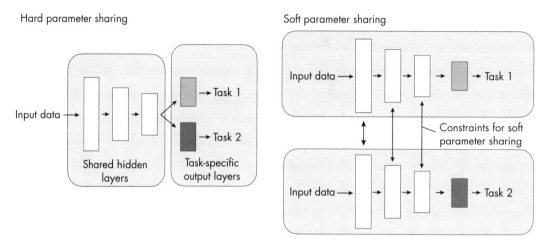

Figure 30-8: The two main types of multi-task learning

In *hard* parameter sharing, as shown in Figure 30-8, only the output layers are task specific, while all the tasks share the same hidden layers and neural network backbone architecture. In contrast, *soft* parameter sharing uses separate neural networks for each task, but regularization techniques such as distance minimization between parameter layers are applied to encourage similarity among the networks.

Multimodal Learning

While multi-task learning involves training a model with multiple tasks and loss functions, multimodal learning focuses on incorporating multiple types of input data.

Common examples of multimodal learning are architectures that take both image and text data as input (though multimodal learning is not restricted to only two modalities and can be used for any number of input modalities). Depending on the task, we may employ a matching loss that forces the embedding vectors between related images and text to be similar, as shown in Figure 30-9. (See Chapter 1 for more on embedding vectors.)

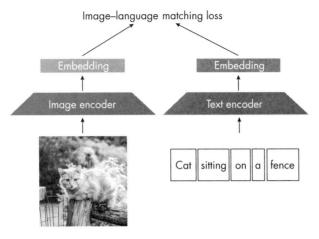

Figure 30-9: Multimodal learning with a matching loss

Figure 30-9 shows image and text encoders as separate components. The image encoder can be a convolutional backbone or a vision transformer, and the language encoder can be a recurrent neural network or language transformer. However, it's common nowadays to use a single transformer-based module that can simultaneously process image and text data. For example, the VideoBERT model has a joint module that processes both video and text for action classification and video captioning.

Optimizing a matching loss, as shown in Figure 30-9, can be useful for learning embeddings that can be applied to various tasks, such as image classification or summarization. However, it is also possible to directly optimize the target loss, like classification or regression, as Figure 30-10 illustrates.

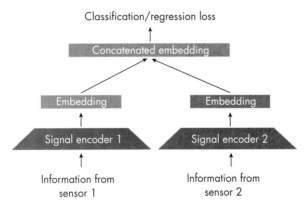

Figure 30-10: Multimodal learning for optimizing a supervised learning objective

Figure 30-10 shows data being collected from two different sensors. One could be a thermometer and the other could be a video camera. The signal encoders convert the information into embeddings (sharing the same number of dimensions), which are then concatenated to form the input representation for the model.

Intuitively, models that combine data from different modalities generally perform better than unimodal models because they can leverage more information. Moreover, recent research suggests that the key to the success of multimodal learning is the improved quality of the latent space representation.

Inductive Biases

Choosing models with stronger inductive biases can help lower data requirements by making assumptions about the structure of the data. For example, due to their inductive biases, convolutional networks require less data than vision transformers, as discussed in Chapter 13.

Recommendations

Of all these techniques for reducing data requirements, how should we decide which ones to use in a given situation?

Techniques like collecting more data, data augmentation, and feature engineering are compatible with all the methods discussed in this chapter. Multi-task learning and multimodal inputs can also be used with the learning strategies outlined here. If the model suffers from overfitting, we should also include techniques discussed in Chapters 5 and 6.

But how can we choose between active learning, few-shot learning, transfer learning, self-supervised learning, semi-supervised learning, and weakly supervised learning? Deciding which supervised learning technique(s) to try is highly context dependent. You can use the diagram in Figure 30-11 as a guide to choosing the best method for your particular project.

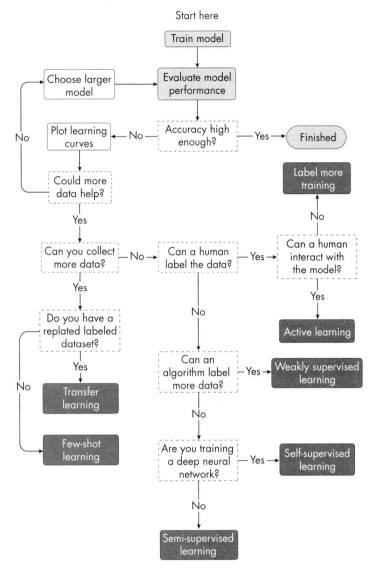

Figure 30-11: Recommendations for choosing a supervised learning technique

Note that the dark boxes in Figure 30-11 are not terminal nodes but arc back to the second box, "Evaluate model performance"; additional arrows were omitted to avoid visual clutter.

Exercises

30-1. Suppose we are given the task of constructing a machine learning model that utilizes images to detect manufacturing defects on the outer shells of tablet devices similar to iPads. We have access to millions of images of various computing devices, including smartphones, tablets, and computers, which are not labeled; thousands of labeled pictures of smartphones depicting various types of damage; and hundreds of labeled images specifically related to the target task of detecting manufacturing defects on tablet devices. How could we approach this problem using self-supervised learning or transfer learning?

30-2. In active learning, selecting difficult examples for human inspection and labeling is often based on confidence scores. Neural networks can provide such scores by using the logistic sigmoid or softmax function in the output layer to calculate class-membership probabilities. However, it is widely recognized that deep neural networks exhibit overconfidence on out-of-distribution data, rendering their use in active learning ineffective. What are some other methods to obtain confidence scores using deep neural networks for active learning?

References

- While decision trees for incremental learning are not commonly implemented, algorithms for training decision trees in an iterative fashion do exist: *https://en.wikipedia.org/wiki/Incremental_decision_tree*.

- Models trained with multi-task learning often outperform models trained on a single task: Rich Caruana, "Multitask Learning" (1997), *https://doi.org/10.1023%2FA%3A1007379606734*.

- A single transformer-based module that can simultaneously process image and text data: Chen Sun et al., "VideoBERT: A Joint Model for Video and Language Representation Learning" (2019), *https://arxiv.org/abs/1904.01766*.

- The aforementioned research suggesting the key to the success of multimodal learning is the improved quality of the latent space representation: Yu Huang et al., "What Makes Multi-Modal Learning Better Than Single (Provably)" (2021), *https://arxiv.org/abs/2106.04538*.

- For more information on active learning: Zhen et al., "A Comparative Survey of Deep Active Learning" (2022), *https://arxiv.org/abs/2203.13450*.

- For a more detailed discussion on how out-of-distribution data can lead to overconfidence in deep neural networks: Anh Nguyen, Jason Yosinski, and Jeff Clune, "Deep Neural Networks Are Easily Fooled: High Confidence Predictions for Unrecognizable Images" (2014), *https://arxiv.org/abs/1412.1897*.

AFTERWORD

This book has taken you on a journey that began with the foundational concepts of machine learning, such as embeddings, latent spaces, and representations, and advanced techniques and architectures, including self-supervised learning, few-shot learning, and transformers. It also covered many practical techniques, such as model deployment, multi-GPU training, and data-centric AI, and dove into specialized domains like computer vision and natural language processing.

Each chapter of this book has not only equipped you with conceptual knowledge, but also offered practical insights, making this book useful for both academic and real-world applications. Whether you're an aspiring data scientist, a machine learning engineer, or just intrigued by the rapidly evolving field of AI, I hope this book has been helpful to you.

If you have found this book valuable, I would be grateful if you could share your experience and spread the word to others who may find it helpful as well. I would also be happy to hear any comments and suggestions. Please feel free to open and engage in discussion on this book's official forum at *https://github.com/rasbt/MachineLearning-QandAI-book/discussions*. You can also see *https://sebastianraschka.com/contact/* for the best ways to get in touch.

Thank you for reading, and I wish you the best of luck in your endeavors in the fascinating world of machine learning and AI.

APPENDIX: ANSWERS TO THE EXERCISES

Chapter 1

1-1. The final layer before the output layer (the second fully connected layer in this case) may be most useful for embeddings. However, we could also use all other intermediate layers to create embeddings. Since the later layers tend to learn higher-level features, these later layers are typically more semantically meaningful and better suited for different types of tasks, including related classification tasks.

1-2. One of the traditional methods of input representation that is different from embeddings is one-hot encoding, as discussed in Chapter 1. In this method, each categorical variable is represented using a binary vector where only one value is "hot" or active (for instance, set to 1), while all other positions remain inactive (for instance, set to 0).

Another representation that is not an embedding is histograms. A typical example of this is image histograms (see *https://en.wikipedia.org/wiki/Image_histogram* for examples). These histograms provide a graphical representation of the tonal distribution in a digital image, capturing the intensity distribution of pixels.

Additionally, the bag-of-words model offers another approach distinct from embeddings. In this model, an input sentence is represented as an unordered collection or "bag" of its words, disregarding grammar and even word order. For more details about the bag-of-words model, see *https://en.wikipedia.org/wiki/Bag-of-words_model*.

Chapter 2

2-1. One way to apply self-supervised learning to video data is by predicting the next frame in the video. This is analogous to next-word prediction in large language models such as GPT. This method challenges the model to anticipate subsequent events or movements in a sequence, giving it a temporal understanding of the content.

Another approach is to predict missing or masked frames. This idea draws inspiration from large language models like BERT, where certain words are masked and the model is tasked with predicting them. In the case of video, entire frames can be masked, and the model learns to interpolate and predict the masked frame based on the context provided by surrounding frames.

Inpainting is yet another avenue for self-supervised learning in videos. Here, instead of masking entire frames, specific pixel areas within a frame are masked. The model is then trained to predict the missing or masked parts, which can help it grasp fine-grained visual details and spatial relationships in the video content.

Lastly, a coloring technique can be used where the video is converted to grayscale and the model is tasked with predicting the color. This not only teaches the model about the original colors of objects, but it also gives insights into lighting, shadows, and the general mood of the scenes.

2-2. We can remove (mask) feature values and train a model to predict these, analogously to classic data imputation. For example, one method that uses this approach is TabNet; see Sercan O. Arik and Tomas Pfister, "TabNet: Attentive Interpretable Tabular Learning" (2019), *https://arxiv.org/abs/1908.07442*.

It is also possible to use contrastive learning by generating augmented versions of the training examples in the original raw feature space or the embedding space. For example, the SAINT and SCARF methods employ this approach. For the former, see Gowthami Somepalli et al., "SAINT: Improved Neural Networks for Tabular Data via Row Attention and Contrastive Pre-Training" (2021), *https://arxiv.org/abs/2106.01342*. For the latter, see Dara Bahri et al., "SCARF: Self-Supervised Contrastive Learning Using Random Feature Corruption" (2021), *https://arxiv.org/abs/2106.15147*.

Chapter 3

3-1. Similar to a supervised learning approach, we first divide the dataset into a training set and a test set. We then further divide the training and test sets into subsets, with one image from each class. To design the training task, we consider only a subset of classes, such as the classes (digits) 0, 1, 2, 5, 6, 8, 9. Next, for testing, we use the remaining classes 3, 4, 7. For each classification task, the neural network receives only one example per image.

3-2. Consider a medical imaging scenario for rare diseases. The training dataset may consist of only a few examples corresponding to different types of diseases, and a few-shot system may have only one or a handful of cases for a new, unseen rare disease (not contained in the training set). The task is then to identify a new rare disease based on this limited number of examples.

Another example of a few-shot system is a recommender that has only a limited number of items a user rated. Based on this limited number of examples, the model has to predict future products the user may like. Imagine a warehouse robot that has to learn to recognize new objects as a company increases its inventory. The robot has to learn to recognize and adapt to these new objects based on only a few examples.

Chapter 4

4-1. You might try increasing the size of the initial neural network. It might be possible that the chosen network is too small to contain a suitable subnetwork.

Another option is to try a different random initialization (for example, by changing the random seed). The lottery hypothesis assumes that *some* randomly initialized networks contain highly accurate subnetworks that can be obtained by pruning, but not all networks may have such subnetworks.

4-2. When training a neural network with ReLU activation functions, specific activations will be set to 0 if the function input is less than 0. This causes certain nodes in the hidden layers not to contribute to the computations; these nodes are sometimes called *dead neurons*. While ReLU activations do not directly cause sparse weights, the zero activation outputs sometimes lead to zero weights that are not recoverable. This observation supports the lottery hypothesis, which suggests that well-trained networks may contain subnetworks with sparse, trainable weights that can be pruned without loss of accuracy.

Chapter 5

5-1. XGBoost is a tree-based gradient-boosting implementation that does not, at the time of writing, support transfer learning. In contrast to artificial neural networks, XGBoost is a nonparametric model that we cannot readily update as new data arrives; hence, regular transfer learning would not work here.

However, it is possible to use the results of an XGBoost model trained on one task as features for another XGBoost model. Consider an overlapping set of features for both datasets. For example, we could design a classification task in a self-supervised fashion for the combined dataset. We could then train a second XGBoost model on the target dataset that takes the original feature set as input, along with the output of the first XGBoost model.

5-2. When applying data augmentations, we usually have to increase the training time as well; it is possible that we needed to train the model for a longer period.

Alternatively, we may have applied too much data augmentation. Augmenting the data too much can result in excessive variations that do not reflect the natural variations in the data, leading to overfitting or poor generalization to new data. In the case of MNIST, this can also include translating or cropping the image in such a way that the digits become unrecognizable due to missing parts.

Another possibility is that we've applied naive, domain-inconsistent augmentation. For example, suppose we are mirroring or flipping images vertically or horizontally. For MNIST, this doesn't make sense, because flipping handwritten digits vertically or horizontally would create numbers that don't exist in the real world.

Chapter 6

6-1. Tuning the number of training epochs is a simpler and more universal approach. This is especially true for older frameworks that don't support model checkpointing. Changing the number of training epochs may therefore be an easier solution and is particularly attractive for small datasets and models where each hyperparameter configuration is cheap to run and evaluate. This approach also eliminates the need for monitoring the performance on a validation set during training, making it straightforward and easy to use.

The early-stopping and checkpointing approach is especially useful when working with models that are expensive to train. It's generally also a more flexible and robust method for preventing overfitting. However, a downside of this approach is that, in noisy training regimes, we may end up prioritizing an early epoch even though the validation set accuracy is not a good estimate of the generalization accuracy.

6-2. One obvious downside of ensemble methods is the increased computational cost. For example, if we build a neural network ensemble of five neural networks, this ensemble can be five times as expensive as every single model.

While we often consider the inferencing costs mentioned above, the increased storage cost is another significant limitation. Nowadays, most computer vision and language models have millions or even billions of parameters that have to be stored in a distributed setting. Model ensembling complicates this further.

Reduced interpretability is yet another cost that we incur when using model ensembles. Understanding and analyzing the predictions of a single model can already be challenging. Depending on the ensembling approach, we add yet another layer of complexity that reduces interpretability.

Chapter 7

7-1. The Adam optimizer implements an adaptive method that comes with internal weight parameters. Adam has two optimizer parameters (mean and variance) per model parameter, so instead of only splitting the weight tensors of the model, we also have to split the optimizer states to work around memory limitations. (Note that this is already implemented in most DeepSpeed parallelization techniques.)

7-2. Data parallelism could theoretically work on a CPU, but the benefits would be limited. For example, instead of duplicating the model in CPU memory to train multiple models on different batches of the dataset in parallel, it could make more sense to increase the data throughput.

Chapter 8

8-1. Self-attention has quadratic compute and memory complexity due to the n-to-n comparisons (where n is the input sequence length), which makes transformers computationally costly compared to other neural network architectures. Moreover, decoder-style transformers such as GPT generate outputs one token at a time, which cannot be parallelized during inference (although generating each token is still highly parallelizable, as discussed in Chapter 8).

8-2. Yes, we can think of self-attention as a form of feature selection, although there are differences between this and other types of feature selection. It is important to differentiate between hard and soft attention in this context. Soft attention computes importance weights for all inputs, whereas hard attention selects a subset of the inputs. Hard attention is more like masking, where certain inputs are set to 0 or 1, while soft attention allows for a continuous range of importance scores. The main difference between attention and feature selection is that feature selection is typically a fixed operation, while attention weights are computed dynamically based on the input. With feature-selection algorithms, the selected features are always the same, whereas with attention, the weights can change based on the input.

Chapter 9

9-1. Automating this evaluation is inherently difficult, and the gold standard is currently based on human evaluation and judgment. However, a few metrics exist as quantitative measures.

   To evaluate the diversity of the generated images, one can compare the conditional class distribution and the marginal class distribution of generated samples, using, for example, a Kullback–Leibler-divergence (KL-divergence) regularization term. This measure is also used in the VAE to make the latent space vectors similar to a standard Gaussian. The higher the KL-divergence term, the more diverse the generated images.

One can also compare the statistics of generated images to real images in the feature space of a pretrained model, such as a convolutional network trained as an image classifier. A high similarity (or low distance) indicates that the two distributions are close to each other, which is generally a sign of better image quality. This approach is also often known as the *Fréchet inception distance approach*.

9-2. Like the generators of GANs, VAEs, or diffusion models, a consistency model takes a noise tensor sampled from a simple distribution (such as a standard Gaussian) as input and generates a new image.

Chapter 10

10-1. Yes, we can make top-k sampling deterministic by setting $k = 1$ so that the model will always select the word with the highest probability score as the next word when generating the output text.

We can also make nucleus sampling deterministic, such as by setting the probability mass threshold p such that it includes only one item, which either exactly meets or exceeds this threshold. This would make the model always choose the token with the highest probability.

10-2. In some cases, the random behavior of dropout during inference can be desirable, such as when building model ensembles with a single model. (Without the random behavior in dropout, the model would produce exactly the same results for a given input, which would make an ensemble redundant.)

Moreover, the random inference behavior in dropout can be useful for robustness testing. For critical applications, like healthcare or autonomous driving, it's essential to understand how slight variations to the model can impact its predictions. By using deterministic dropout patterns, we can simulate these slight variations and test the robustness of the model.

Chapter 11

11-1. SGD has only the learning rate as a hyperparameter, but it does not have any parameters. Therefore, it does not add any additional parameters to be stored besides the gradients calculated for each weight parameter during backpropagation (including the layer activations required for calculating the gradients).

The Adam optimizer is more complex and requires more storage. Specifically, Adam keeps an exponentially decaying average of past gradients (first moment) and an exponentially decaying average of past squared gradients (second raw moment) for each parameter. Therefore, for each parameter in the network, Adam needs to store two additional values. If we have n parameters in the network, Adam requires storage for $2n$ additional parameters.

If the network has n trainable parameters, Adam adds $2n$ parameters to be tracked. For example, in the case of AlexNet, which consists of 26,926 parameters, as calculated in Exercise 1-1, Adam requires 53,852 additional values in total ($2 \times 26{,}926$).

11-2. Each BatchNorm layer learns two sets of parameters during training: a set of scaling coefficients (gamma) and a set of shifting coefficients (beta). These are learned so that the model can undo the normalization when it is found to be detrimental to learning. Each of these sets of parameters (gamma and beta) has the same size as the number of channels (or neurons) in the layer they normalize because these parameters are learned separately for each channel (or neuron).

For the first BatchNorm layer following the first convolutional layer with five output channels, this adds 10 additional parameters. For the second BatchNorm layer, following the second convolutional layer with 12 output channels, this adds 24 additional parameters.

The first fully connected layer has 128 output channels, which means 256 additional BatchNorm parameters. The second fully connected layer is not accompanied by a BatchNorm layer since it's the output layer.

Therefore, BatchNorm adds 10 + 24 + 256 = 290 additional parameters to the network.

Chapter 12

12-1. Just increasing the stride from 1 to 2 (or larger values) should not affect the equivalence since the kernel size is equal to the input size in both scenarios, so there is no sliding window mechanism at play here.

12-2. Increasing the padding to values larger than 0 will affect the results. Due to the padded inputs, we will have the sliding window convolutional operation where the equivalence with fully connected layers no longer holds. In other words, the padding would alter the input's spatial dimensions, which would no longer match the kernel size and would result in more than one output value per feature map.

Chapter 13

13-1. Using smaller patches increases the number of patches for a given input image, leading to a higher number of tokens being fed into the transformer. This results in increased computational complexity, as the self-attention mechanism in transformers has quadratic complexity with respect to the number of input tokens. Consequently, smaller input patches make the model computationally more expensive.

13-2. Using larger input patches may result in the loss of finer details and local structures in the input image, which can potentially negatively affect the model's predictive performance. Interested readers might enjoy the FlexiViT paper that studies the computational and predictive performance trade-offs as a consequence of the patch size and number (Lucas Beyer et al., "FlexiViT: One Model for All Patch Sizes" [2022], *https://arxiv.org/abs/2212.08013*).

Chapter 14

14-1. Because homophones have different meanings, we expect them to appear in other contexts, such as *there* and *their* in "I can see you over there" and "Their paper is very nice."

Since the distributional hypothesis says that words with similar meanings should appear in similar contexts, homophones do not contradict the distributional hypothesis.

14-2. The underlying idea of the distributional hypothesis can be applied to other domains, such as computer vision. In the case of images, objects that appear in similar visual contexts are likely to be semantically related. On a lower level, neighboring pixels are likely semantically related, as they are part of the same object; this idea is used in masked autoencoding for self-supervised learning on image data. (We covered masked autoencoders in Chapter 2.)

Another example is protein modeling. For example, researchers showed that language transformers trained on protein sequences (a string representation where each letter represents an amino acid, like *MNGTEGPNFYVPFSNKTGVV . . .*) learn embeddings where similar amino acids cluster together (Alexander Rives et al., "Biological Structure and Function Emerge from Scaling Unsupervised Learning to 250 Million Protein Sequences" [2019], *https://www.biorxiv.org/content/10 .1101/622803v1.full*). The hydrophobic amino acids such as V, I, L, and M appear in one cluster, and aromatic amino acids such as F, W, and Y appear in another cluster. In this context, we can think of an amino acid as an equivalent to a word in a sentence.

Chapter 15

15-1. Assuming that the existing data does not suffer from privacy concerns, data augmentation helps generate variations of the existing data without the need to collect additional data, which can help with privacy concerns.

However, if the original data includes personally identifiable information, even augmented or synthetic data could potentially be linked back to individuals, especially if the augmentation process doesn't sufficiently obscure or alter the original data.

15-2. Data augmentation might be less beneficial if the original dataset is already large and diverse enough that the model isn't overfitting or underperforming due to a lack of data. This is, for example, often the case when pretraining LLMs. The performance of highly domain-specific models (for example, in the medical, law, and financial domains) could also be adversely affected by techniques such as synonym replacement and back translation due to replacing domain-specific terms with a certain meaning. In general, in contexts of tasks highly sensitive to wording choices, data augmentation must be applied with particular care.

Chapter 16

16-1. The self-attention mechanism has quadratic time and memory complexity. More precisely, we can express the time and memory complexity of self-attention as $O(N^2 \times d)$, where N is the length of the sequence and d is the dimensionality of the embedding of each element in the sequence.

This is because self-attention involves computing a similarity score between each pair of elements in the sequence. For example, we have an input matrix X with N tokens (rows) where each token is a d-dimensional embedding (columns).

When we compute the dot product of each token embedding to each other token embedding, we multiply XX^T, which results in an $N \times N$ similarity matrix. This multiplication involves d multiplications for a single token pair, and we have N^2 such pairs. Hence, we have $O(N^2 \times d)$ complexity. The $N \times N$ similarity matrix is then used to compute weighted averages of the sequence elements, resulting in an $N \times d$ output representation. This can make self-attention computationally expensive and memory intensive, particularly for long sequences or large values of d.

16-2. Yes. Interestingly, self-attention may partly be inspired by the spatial attention mechanisms used in convolutional neural networks for image processing (Kelvin Xu et al., "Show, Attend and Tell: Neural Image Caption Generation with Visual Attention" [2015], *https://arxiv.org/abs/1502.03044*). Spatial attention is a mechanism that allows a neural network to focus on specific regions of an image that are relevant to a given task. It works by selectively weighting the importance of different spatial locations in the image, which allows the network to "pay more attention" to certain areas and ignore others.

Chapter 17

17-1. To adapt a pretrained BERT model for classification, you need to add an output layer for classification, often referred to as a *classification head*.

As discussed, BERT uses a [CLS] token for the next-sentence prediction task during pretraining. Instead of training it for next-sentence prediction, we can fine-tune a new output layer for our target prediction task, such as sentiment classification.

The embedded [CLS] output vector serves as a summary of the entire input sequence. We can think of it as a feature vector and train a small neural network on top of it, typically a fully connected layer followed by a softmax activation function to predict the class probabilities. The fully connected layer's output size should match the number of classes in our classification task. Then we can train it using back-propagation as usual. Different fine-tuning strategies (updating all layers versus only the last layer) can then be used to train the model on a supervised dataset, for example.

17-2. Yes, we can fine-tune a decoder-only model like GPT for classification tasks, although it might not be as effective as using encoder-based models like BERT. In contrast to BERT, we do not need to use a special [CLS], but the fundamental concept is similar to fine-tuning an encoder-style model for classification. We add a classification head (a fully connected layer and a softmax activation) and train it on the embedding (the final hidden state) of the first generated output token. (This is analogous to using the [CLS] token embedding.)

Chapter 18

18-1. In-context learning is useful if we don't have access to the model or if we want to adapt the model to similar tasks that the model wasn't trained to do.

In contrast, fine-tuning is useful for adapting the model to a new target domain. For example, suppose the model was pretrained on a general corpus and we want to apply it to financial data or documents. Here, it would make sense to fine-tune the model on data from that target domain.

Note that in-context learning can be used with a fine-tuned model as well. For example, when a pretrained language model is fine-tuned on a specific task or domain, in-context learning then leverages the model's ability to generate responses based on the context provided within the input that may be more accurate given the target domain compared to in-context learning without fine-tuning.

18-2. This is done implicitly. In prefix tuning, adapters, and LoRA, the original knowledge of the pretrained language model is preserved by keeping the core model parameters frozen while introducing additional learnable parameters that adapt to the new task.

Chapter 19

19-1. If we were using an embedding technique like Word2Vec that processes each word independently, we would expect the cosine similarity between the "cat" embeddings to be 1.0. However, we are using a transformer model to produce the embeddings in this case. Transformers use self-attention mechanisms that consider the whole context (for instance, input text) when producing the embedding vectors. (See Chapter 16 for more information about self-attention.) Since the word *cat* is used in two different sentences, the BERT model produces a different embedding for these two instances of *cat*."

19-2. Switching the candidate and reference texts has the same effect as calculating the maximum cosine similarity scores across columns (as shown in step 5 of Figure 19-3) versus rows, which can result in different BERTScores for specific texts. That's why the BERTScore is often computed as an F1 score similar to ROUGE in practice. For instance, we calculate the BERTScore one way (recall), then the other (precision), and then compute the harmonic mean (F1 score).

Chapter 20

20-1. Random forests, typically based on CART decision trees, cannot be readily updated as new data arrives. Hence, a stateless training approach would be the only viable option. On the other hand, suppose we switched to using neural network models such as recurrent neural networks. In that case, a stateful approach could make more sense since the neural network could be readily updated on new data. (However, in the beginning, comparing stateful and stateless systems side by side is always a good idea before deciding which method works best.)

20-2. A stateful retraining approach makes the most sense here. Instead of training a new model on a combination of existing data, including user feedback, it makes more sense to update the model based on user feedback. Large language models are usually pretrained in a self-supervised fashion and then fine-tuned via supervised learning. Training large language models is very expensive, so updating the model via stateful retraining makes sense rather than training it from scratch again.

Chapter 21

21-1. From the information provided, it is unclear whether this is a data-centric approach. The AI system relies heavily on data inputs to make predictions and recommendations, but that's true for any machine learning approach for AI. To determine whether this approach is an example of data-centric AI, we need to know how the AI system was developed. If it was developed by using a fixed model and refining the training data, this could qualify as a data-centric approach; otherwise, it's just regular machine learning and predictive modeling.

21-2. If we are keeping the model fixed—that is, reusing the same ResNet-34 architecture—and are changing only the data augmentation approach to investigate its influence on the model performance, we could consider this a data-centric approach. However, data augmentation is also routinely done as part of any modern machine learning pipeline, and the use of data augmentation alone does not tell us whether an approach is data centric. Under the modern definition, a data-centric approach entails actively studying the difference between various dataset-enhancing techniques while keeping the remaining modeling and training pipeline fixed.

Chapter 22

22-1. One downside of using multi-GPU strategies for inference is the additional communication overhead between the GPUs. However, for inference tasks, which are relatively small compared to training since they don't require gradient computations and updates, the time it takes to communicate between GPUs could outweigh the time saved by parallelization.

Managing multiple GPUs also means higher equipment and energy costs. In practice, optimizing models for single-GPU or CPU performance is usually more worthwhile. If multiple GPUs are available, processing multiple samples in parallel on separate GPUs often makes more sense than processing the same sample via multiple GPUs.

22-2. Loop tiling is often combined with vectorization. For example, after applying loop tiling, each tile can be processed using vectorized operations. This allows us to use SIMD instructions on data that is already in the cache, increasing the effectiveness of both techniques.

Chapter 23

23-1. The problem is that importance weighting assumes the test set distribution matches the deployment distribution. However, this is often not the case for various reasons, such as changing user behavior, evolving product features, or dynamic environments.

23-2. It's common to monitor metrics such as classification accuracy, where a drop in performance may indicate a shift in the data. However, this is impractical if we don't have access to the labels of incoming data.

In cases where it's infeasible to label new, incoming data, we can use statistical two-sample tests to determine whether the examples come from the same distribution. We can also use adversarial validation, discussed in Chapter 29. However, these methods won't help detect concept shifts, as they compare only input distributions, not the relationship between inputs and outputs.

Other methods include measuring the reconstruction error: if we have an autoencoder trained on our source data, we can monitor the reconstruction error on new data. If the error increases significantly, it may indicate a shift in the input distribution.

Outlier detection is another common technique. Here, unusually high rates of data points being identified as outliers could suggest a shift in data distribution.

Chapter 24

24-1. Trying to predict the number of goals a player scores (based on data from past seasons, for example) is a Poisson regression problem. On the other hand, we could also apply an ordinal regression model to the different players to rank them by the number of goals they will score.

However, since the goal difference is constant and can be quantified (for example, the difference between 3 and 4 goals is the same as the difference between 15 and 16 goals), it's not an ideal problem for an ordinal regression model.

24-2. This is a ranking issue that resembles an ordinal regression issue, but there are some differences. Since we are aware of only the relative order of the movies, a pairwise ranking algorithm might be a more appropriate solution than an ordinal regression model.

However, if the person is asked to assign numerical labels to each movie on a scale such as 1 to 5 (similar to the star rating system on Amazon), it would be possible to train and use an ordinal regression model on this type of data.

Chapter 25

25-1. The choice of the confidence level (90 percent, 95 percent, 99 percent, and so forth) affects the width of the confidence interval. A higher confidence level will produce a wider interval because we need to cast a wider net to be more confident that we have captured the true parameter.

Conversely, a lower confidence level produces a narrower interval, reflecting more uncertainty about where the true parameter lies. A 90 percent confidence interval is therefore narrower than a 95 percent confidence interval, reflecting greater uncertainty about the location of the true population parameter. Colloquially speaking, we are 90 percent certain that the true parameter lies within a small range of values. To increase this certainty, we must increase the width to 95 percent or 99 percent.

For example, let's say we are 90 percent certain that it will rain in the next two weeks in Wisconsin. If we want to make a prediction with 95 percent confidence without collecting additional data, we would have to increase the time interval. For example, we might say that we are 95 percent certain that it will rain within the next four weeks, or 99 percent certain that it will rain within the next two months.

25-2. Since the model is already trained and stays the same, applying it to each test set would be wasteful. To speed up the process outlined in this section, we technically need to apply the model only once, namely on the original test set. We can then bootstrap the actual and predicted labels directly (instead of the original samples) to create the bootstrapped test sets. We can then compute the test set accuracies based on the bootstrapped labels in each set.

Chapter 26

26-1. The prediction set size can tell us a lot about the certainty of the prediction. If the prediction set size is small (for example, 1 in classification tasks), it indicates a high level of confidence in the prediction. The algorithm has enough evidence to strongly suggest one specific outcome.

If the prediction set size is larger (for example, 3 in classification tasks), it indicates more uncertainty. The model is less confident about the prediction and considers multiple outcomes to be plausible. In practice, we can use this information to assign more resources to examples with a high prediction set size. For example, we may flag these cases for human verification since the machine learning model is less certain.

26-2. Absolutely. Confidence intervals are just as applicable to regression models as they are to classification models. In fact, they're even more versatile in the context of regression. For example, we can compute confidence intervals for the performance of a model, like the mean squared error, using the methods illustrated in Chapter 25. (But we can also compute confidence intervals for individual predictions and model parameters. If you're interested in confidence intervals for model parameters, check out my article "Interpretable Machine Learning—Book Review and Thoughts About Linear and Logistic Regression as Interpretable Models" at *https://sebastianraschka.com/blog/2020/interpretable -ml-1.html*.)

We can also compute conformal prediction intervals for regression models. The interval is a range of possible target values instead of a single point estimate. The interpretation of such a prediction interval is that, under the assumption that the future is statistically similar to the past (for instance, based on the data the model was trained on), the true target value for a new instance will fall within this range with a certain confidence level, such as 95 percent.

Chapter 27

27-1. Since the MAE is based on an absolute value around the distance, it naturally satisfies the first criterion: it can't be negative. Also, the MAE is the same if we swap the values y and $\hat{y}$; hence, it satisfies the second criterion. But how about the triangle inequality? Similar to how the RMSE is the same as the Euclidean distance or L2 norm, the MAE is similar to the L1 norm between two vectors. Since all vector norms satisfy the triangle inequality (Horn and Johnson, *Matrix Analysis*, Cambridge University Press, 1990), our colleague is incorrect.

Furthermore, even if the MAE were not a proper metric, it could still be a useful model evaluation metric; for example, consider the classification accuracy.

27-2. The MAE assigns equal weight to all errors, while the RMSE places more emphasis on errors with larger absolute values due to the quadratic exponent. As a result, the RMSE is always at least as large as the MAE. However, no metric is universally better than the other, and they have both been used to assess model performance in countless studies over the years.

If you are interested in additional comparisons between MAE and RMSE, you may like the article by Cort J. Willmott and Kenji Matsuura, "Advantages of the Mean Absolute Error (MAE) Over the Root Mean Square Error (RMSE) in Assessing Average Model Performance" (2005), *https://www.int-res.com/abstracts/cr/v30/n1/p79-82.*

Chapter 28

28-1. This is not a problem if we care about only the average performance. For example, if we have a dataset of 100 training examples, and the model predicts 70 out of the 100 validation folds correctly, we estimate the model accuracy as 70 percent. However, suppose we are interested in analyzing the variance of the estimates from the different folds. In that case, LOOCV is not very useful: since each fold consists of only a single training example, we cannot compute the variance of each fold and compare it to other folds.

28-2. Another use case of *k*-fold cross-validation is model ensembling. For example, in 5-fold cross-validation, we train five different models since we have five slightly different training sets. However, instead of training a final model on the whole training set, we can combine the five models into a model ensemble (this is particularly popular on Kaggle). See Figure 6-3 on page 34 for an illustration of this process.

Chapter 29

29-1. As a performance baseline, it's a good idea to implement a zero-rule classifier, such as a majority class classifier. Since we typically have more training data than test data, we can compute the performance of a model that always predicts *Is test? False*, which should result in 70 percent accuracy if we have partitioned the original dataset into 70 percent training data and 30 percent test data. If the accuracy of the model trained on the adversarial validation dataset noticeably exceeds this baseline (say, 80 percent), we may have a serious discrepancy issue to investigate further.

29-2. Overall, this is not a big issue since we are mainly interested in whether there is a strong deviation from a majority class prediction baseline. For instance, if we compare the accuracy of the adversarial validation model against the baseline (rather than 50 percent accuracy), there should be no issue. However, it may be even better to consider evaluation metrics like Matthew's correlation coefficient or ROC or precision-recall area-under-the-curve values instead of classification accuracy.

Chapter 30

30-1. While we often think of self-supervised learning and transfer learning as separate approaches, they don't have to be exclusive. For instance, we could pretrain a model on a labeled or larger unlabeled image dataset using self-supervised learning (in this case, the millions of unlabeled images corresponding to the various computing devices).

Instead of starting with random weights, we can use the neural network weights from self-supervised learning to follow up with transfer learning via the thousands of labeled smartphone pictures. Since smartphones are related to tablets, transfer learning is a very promising approach here.

Finally, after the self-supervised pretraining and transfer learning, we can fine-tune the model on the hundreds of labeled images of the target task, the tablets.

30-2. Besides mitigation techniques for the overconfident scores from a neural network's output layer, we can also consider various ways of ensembling to obtain confidence scores. For instance, instead of disabling dropout during inference, we can leverage dropout to obtain multiple different predictions for a single example to compute the predicted label uncertainty.

Another option is to construct model ensembles from different segments of the training set using k-fold cross-validation, as discussed in the ensemble section of Chapter 6.

It is also possible to apply conformal prediction methods, discussed in Chapter 26, to active learning.

INDEX

leave-one-out cross-validation (LOOCV), 188, 221
10-fold cross-validation, 187, 188
CUDA Deep Neural Network library (cuDNN), 62

D

data. *See also* limited labeled data
    applying self-supervised learning to video, 14, 208
    count, 161
    reducing overfitting with, 23–27, 209–210
    self-supervised learning for tabular, 14, 208
    synthetic, generation of, 96–97
    unlabeled, in self-supervised learning, 10, 11
data augmentation
    to reduce overfitting, 24–25, 26, 210
    for text, 93–97, 214–215
data-centric AI, 143–146, 217
data distribution shifts
    concept drift, 155
    covariate shift, 153–154
    domain shift, 155–156
    label shift, 154–155
    overview, 153
    types of, 156–157
data parallelism, 37, 38, 39–40, 41–42, 211
datasets
    for few-shot learning, 15
    sampling and shuffling as source of randomness, 60
    for transformers, 45
DBMs (deep Boltzmann machines), 50–51, 57
dead neurons, 209
decision trees, 204
decoder network (VAE model), 51–52
decoders
    in Bahdanau attention mechanism, 100–101
    in original transformer architecture, 105–106, 107

decoder-style transformers. *See also* encoder-style transformers
    contemporary transformer models, 111–112
    distributional hypothesis, 91
    encoder-decoder hybrids, 110
    overview, 105, 108–110
    synthetic data generation, 96–97
    terminology related to, 110
deep Boltzmann machines (DBMs), 50–51, 57
deep generative models. *See* generative AI models
deep learning. *See also* generative AI models
    embeddings, 3–7, 207
    few-shot learning, 15–18, 208–209
    lottery ticket hypothesis, 19–21, 209
    multi-GPU training paradigms, 37–42, 211
    reducing overfitting
        with data, 23–27, 209–210
        with model modifications, 29–36, 210
    self-supervised learning, 9–14, 208
    sources of randomness, 59–65, 212
    transformers, success of, 43–47, 211
DeepSpeed, 37, 42
deletion, word, as data augmentation technique, 94
deterministic algorithms, 62, 65
diffusion models, 55–56, 57, 58
dimension contrastive self-supervised learning, 14
direct convolution, 61
discriminative models, 49–50
discriminator in GANs, 52–53
distance, embeddings as encoding, 5
distance functions, 179–183
distributional hypothesis, 89–92, 214
domain shift (joint distribution shift), 155–156, 157
double descent, 32–33, 36
downstream model for pretrained transformers, 114
downstream task, 11
drivers as source of randomness, 62
dropout, 30, 36, 61, 64–65, 212

generative large language models. *See* evaluation metrics for generative LLMs; large language models; natural language processing

generator in GANs, 52–53

Gibbs sampling, 51

GPT (generative pretrained transformer) models
    decoder-style transformers, 91, 109–110
    fine-tuning for classification, 112, 216
    randomness by design, 63
    self-prediction, 12

GPUs. *See* multi-GPU training paradigms

grokking, 32–33, 36

H

hard attention, 211

hard parameter sharing, 200

hard prompt tuning, 117–118

hardware as source of randomness, 62

hierarchical processing in CNNs, 80

histograms, 207

holdout validation as source of randomness, 60

homophones, 92, 214

human feedback, reinforcement learning with, 124

hyperparameter tuning, 188

I

image denoising, 56–57

image generation, 51, 52, 54–57, 211–212

image histograms, 207

"An Image Is Worth 16x16 Words" (Dosovitskiy et al.), 83, 85

ImageNet dataset, 9, 14, 175

image processing. *See* computer vision

importance weighting, 154, 155, 157, 218

in-context learning, 113, 116–119, 125, 216. *See also* few-shot learning

indexing, 118–119, 125

inductive biases
    in convolutional neural networks, 80–82
    limited labeled data, 202
    overview, 79
    in vision transformers, 83–84

inference, speeding up. *See* model inference, speeding up

inpainting, 194–195, 208

input channels in convolutional layers, 70–71, 76–77

input embedding, 4

input representations, 6, 207

InstructGPT model, 124, 126, 133, 135

inter-op parallelism (model parallelism), 37, 38, 39–40, 41–42

intra-op parallelism (tensor parallelism), 37, 38–40, 41–42, 211

intrinsic metrics, 128

iterative pruning, 20, 31

J

joint distribution shift (domain shift), 155–156, 157

K

kernel size in convolutional layers, 70–71, 76–77

k-fold cross-validation
    determining appropriate values for k, 187–188
    ensemble approach, 33–34
    overview, 185–186
    as source of randomness, 60
    trade-offs in selecting values for k, 186–187

knowledge distillation, 31–33, 35, 36, 151, 199

Kullback–Leibler divergence (KL divergence), 32, 52, 211

L

$L2$ distance, 181

$L2$ regularization, 30, 35

labeled data, limited. *See* limited labeled data

label shift (prior probability shift), 154–155, 156

label smoothing, 27

language transformers. *See* transformers

large language models (LLMs). *See also* natural language processing; transformers
distributional hypothesis, 91
evaluation metrics for, 127–135, 216–217
stateless vs. stateful training, 141, 217
synthetic data generation, 96–97

latent space, 3, 5–7

layer input normalization techniques, 34–35

layers
convolutional layers
calculating number of parameters in, 70–71
as high-pass and low-pass filters, 84
recommendations for, 78
replacing fully connected layers with, 75–78
normalization in original transformer architecture, 106–107
updating when fine-tuning pretrained transformers, 115–116
using to create embeddings, 207

leave-one-out cross-validation (LOOCV), 188, 221

limited labeled data
active learning, 195
bootstrapping data, 194
few-shot learning, 195–196
inductive biases, 202
labeling more data, 193–194
meta-learning, 196–197
multimodal learning, 200–202
multi-task learning, 199–200
overview, 193
recommendations for choosing technique, 202–203

self-supervised learning, 194–195
self-training, 199
semi-supervised learning, 198–199
transfer learning, 194
weakly supervised learning, 197–198

linear classifiers, 114

LLMs. *See* large language models; natural language processing; transformers

local connectivity in CNNs, 80, 81

logistic regression classifier, 49–50

LOOCV (leave-one-out cross-validation), 188, 221

loop fusion (operator fusion), 150–151

loop tiling (loop nest optimization), 149–150, 151, 152, 218

LoRA (low-rank adaptation), 119, 123–124, 125, 126, 216

loss function, VAEs, 52

lottery ticket hypothesis
overview, 19
practical implications and limitations, 20–21
training procedure for, 19–20

low-rank adaptation (LoRA), 119, 123–124, 125, 126, 216

low-rank transformation, 123

M

MAE (mean absolute error), 183, 220–221

majority voting, 33

MAPIE library, 178

masked (missing) input self-prediction methods, 12

masked frames, predicting, 208

masked language modeling, 91, 107–108, 194

mean absolute error (MAE), 183, 220–221

mean squared error (MSE) loss, 180–181

memory complexity of self-attention, 103, 215

metadata (meta-features) extraction, 197

meta-learning, 17, 196–197

METEOR metric, 131, 134

metrics, proper. *See* proper metrics

seeding random generator, 60, 61
self-attention mechanism.
   See also transformers
  vs. Bahdanau attention
    mechanism, 99–101
  overview, 99, 101–102
  sequence parallelism, 40
  transformers, 42, 43–45, 46, 47
  in vision transformers, 83–84
self-prediction, 11–12
self-supervised learning
  contrastive, 12–14
  encoder-only architectures, 108
  leveraging unlabeled data, 11
  limited labeled data, 194–195, 203,
    204, 221–222
  overview, 9
  pretraining transformers via, 45
  reducing overfitting with, 25
  self-prediction, 11–14
  vs. transfer learning, 9–11
self-training, 199. *See also* knowledge
   distillation
semi-supervised learning, 198–199, 203
sentence shuffling, 95
[SEP] token, 108
sequence parallelism, 40–41, 42
sequence-to-sequence (seq2seq)
   models, 107–110
sequential inference, 148
SGD (stochastic gradient descent)
   optimizer, 73, 212
shortcut connection, 107
siamese network setup, 13
similarity, embeddings as encoding, 5
.632 bootstrap, 171
skip connection in transformer
   architecture, 107
skip-gram approach, Word2vec, 90
smaller models, reducing overfitting
   with, 31–33
soft attention, 211
soft parameter sharing, 200
soft prompting, 119–121, 125
sources of randomness
  dataset sampling and shuffling, 60
  different runtime algorithms, 61–62
  and generative AI, 62–64

hardware and drivers, 62
  model weight initialization, 59–60
  nondeterministic algorithms, 61
  overview, 59
spatial attention, 215
spatial invariance, 80–82
speeding up inference. *See* model
   inference, speeding up
squared error (SE) loss, 181
Stable Diffusion latent diffusion
   model, 58
stacking (stacked generalization), 33
stateful training, 139, 140–141, 217
stateless training (stateless retraining),
   139–140, 141, 217
statistical population, 164
statistical two-sample tests, 218
stochastic diffusion process, 56
stochastic gradient descent (SGD)
   optimizer, 73, 212
stride, 78, 213
structured pruning, 20
student in knowledge distillation,
   31–32
supervised learning, 15. *See also* limited
   labeled data
support set in few-shot learning, 16
synonym replacement (text
   augmentation), 93–94
synthetic data generation, 96–97

T

TabNet, 208
tabular data, self-supervised learning
   for, 14, 208
teacher in knowledge distillation,
   31–32
10-fold cross-validation, 187, 188
TensorFlow framework, 59, 62, 149
tensor parallelism, 37, 38–40, 41–42, 211
test sets
  bootstrapping, 169, 170, 171
  conformal predictions, 176
  discordance with training sets,
   189–191, 221
text, data augmentation for, 93–97,
   214–215
T5 encoder-decoder architecture, 112

time complexity of self-attention, 103, 215

top-*k* sampling, 63–64, 212

training. *See also* multi-GPU training paradigms; pretraining; randomness, sources of; retraining

    epochs, tuning number of, 35, 210

    post-training quantization, 151

    procedure for lottery ticket hypothesis, 19–20

    quantization-aware, 151

    self-training, 199

    stateless and stateful, 139–141, 217

training sets

    conformal predictions, 176, 177

    discordance with test sets, 189–191, 221

    for vision transformers, 79–85, 213–214

transfer learning

    limited labeled data, 194, 203, 204, 221–222

    reducing overfitting with, 25, 26, 209

    vs. self-supervised learning, 9–11

transformers. *See also* self-attention mechanism

    adapting pretrained language models, 124–125

    attention mechanism, 40, 43–45

    classification tasks, 113–116

    contemporary models, 111–112

    decoders, 108–110

    encoder-decoder hybrids, 110

    encoders, 107–108

    in-context learning, indexing, and prompt tuning, 116–119

    multi-GPU training paradigms, 40, 42

    number of parameters, 45

    original architecture for, 105–110

    overview, 105, 113

    parallelization, 45–46

    parameter-efficient fine-tuning, 119–124

pretraining via self-supervised learning, 45

reinforcement learning with human feedback (RLHF), 124

success of, 43–47

terminology, 110

transfer learning, 11

translation

    back, 96

    invariance and equivariance, 80–82

    tokens, 44–46, 63–64, 102, 106–110, 117–118

triangle inequality, 180, 181, 182

true generalization accuracy, 164

two-dimensional embeddings, 4–5

typo introduction, 95

U

unlabeled data in self-supervised learning, 10, 11

unstructured pruning, 20

unsupervised pretraining. *See* self-supervised learning

V

variational autoencoders (VAEs), 51–52, 53, 54, 57, 58

variational inference, 51

vectorization, 148–149, 152, 218

VideoBERT model, 201, 204

video data, applying self-supervised learning to, 14, 208

vision transformers (ViTs)

    vs. convolutional neural networks, 79, 82–83, 84

    inductive biases in, 83–84

    large training sets for, 79

    positional information in, 82, 85

    recommendations for, 84

W

weakly supervised learning, 197–199, 203

weight decay, 30, 35

weighted loss function, 155

weight initialization, 59–60

weight normalization, 34–35

Never before has the world relied so heavily on the Internet to stay connected and informed. That makes the Electronic Frontier Foundation's mission—to ensure that technology supports freedom, justice, and innovation for all people—more urgent than ever.

For over 30 years, EFF has fought for tech users through activism, in the courts, and by developing software to overcome obstacles to your privacy, security, and free expression. This dedication empowers all of us through darkness. With your help we can navigate toward a brighter digital future.

ELECTRONIC FRONTIER FOUNDATION EFF